Intelligent Autonomous Drones with Cognitive Deep Learning

Build AI-Enabled Land Drones with the Raspberry Pi 4

David Allen Blubaugh
Steven D. Harbour
Benjamin Sears
Michael J. Findler

Apress®

Intelligent Autonomous Drones with Cognitive Deep Learning: Build AI-Enabled Land Drones with the Raspberry Pi 4

David Allen Blubaugh
Springboro, OH, USA

Steven D. Harbour
Beavercreek Township, OH, USA

Benjamin Sears
Xenia, OH, USA

Michael J. Findler
Mesa, AZ, USA

ISBN-13 (pbk): 978-1-4842-6802-5
https://doi.org/10.1007/978-1-4842-6803-2

ISBN-13 (electronic): 978-1-4842-6803-2

Managing Director, Apress Media LLC: Welmoed Spahr
Acquisitions Editor: Susan McDermott
Development Editor: James Markham
Coordinating Editor: Jessica Vakili
Copy Editor: April Rondeau

Distributed to the book trade worldwide by Springer Science + Business Media New York, 233 Spring Street, 6th Floor, New York, NY 10013. Phone 1-800-SPRINGER, fax (201) 348-4505, email orders-ny@springer-sbm.com, or visit www.springeronline.com. Apress Media, LLC is a California LLC and the sole member (owner) is Springer Science + Business Media Finance Inc (SSBM Finance Inc). SSBM Finance Inc is a **Delaware** corporation.

For information on translations, please e-mail booktranslations@springernature.com; for reprint, paperback, or audio rights, please e-mail bookpermissions@springernature.com.

Apress titles may be purchased in bulk for academic, corporate, or promotional use. eBook versions and licenses are also available for most titles. For more information, reference our Print and eBook Bulk Sales web page at http://www.apress.com/bulk-sales.

Any source code or other supplementary material referenced by the author in this book is available to readers on the Github repository: https://github.com/Apress/Intelligent-Autonomous-Drones-with-Cognitive-Deep-Learning. For more detailed information, please visit http://www.apress.com/source-code.

Printed on acid-free paper

Table of Contents

About the Authors

David Allen Blubaugh is an experienced computer and electrical engineer with both bachelor's and master's degrees from Wright State University. He is currently working for ATR, LLC, which is a company located in the Springboro, Ohio, area. At present, he is in the process of completing his UAS drone operator degree program with Benjamin Sears at Sinclair College. He has experience with embedded systems such as the MSP430 microcontroller and the Raspberry Pi 4.

Steven D. Harbour, **PhD**, is staff engineer and scientist in the Dayton Engineering Advanced Projects Lab at the Southwest Research Institute. He is a senior leader and defense research and engineering professional with over 25 years of experience in multiple engineering and aviation disciplines and applications. He leads and performs ongoing basic and applied research projects, including the development of third-generation spiking neural networks (SNNs), neuromorphic engineering, and neuromorphic applications that include human autonomy teaming. He has supported the Air Force Research Laboratory Sensors Directorate at Wright-Patterson Air Force Base in Ohio, and at the Air Force Life Cycle Management Center. He is a USAF test pilot with over 5,000 hours total flying time. He has a PhD in neuroscience, MS in aerospace engineering and mathematics, and BS in electrical engineering. Dr. Harbour also teaches at the University of Dayton and Sinclair College.

ABOUT THE AUTHORS

Benjamin Sears has an in-depth understanding of the theory behind drone missions and crew resource management. He also has applied experience as a drone pilot/operator who conducted missions as a civilian contractor in both the Iraq and Afghanistan areas of operation.

Michael J. Findler is a computer science instructor at Wright State University with experience in embedded systems development. He also has developed and worked in various fields within the universe of artificial intelligence.

CHAPTER 1

Rover Platform Overview

Imagine this: You are an aspiring engineer and founder of a company called Advanced Technologies & Resources, LLC (ATR). Your company has just won a multi-million-dollar contract with the Egyptian government and Egyptian Supreme Council of Antiquities. The Egyptian government wants to explore the inner areas of the pyramids. This includes the caverns, wells, and pits found within the Great Pyramids at Giza. However, there is problem! Some of the caverns and wells are prone to sudden cave-ins, and there is the possibility of undetected "booby-traps" that were set by the pyramid builders themselves back in the year 2553 BC. Furthermore, poisonous gasses, such as carbon monoxide, have been accumulating within the tombs of the pyramids for more than 4,500 years, making these areas dangerous for human exploration.

The Egyptian government wants to explore these areas without sending any human explorers or archaeologists into these unsafe and potentially deadly caverns or wells. However, they cannot send in standard robots, because the attached wires may cause irreparable damage to the interior structure and any artifacts. They also cannot send in wireless robots, since the radio- and data-links degrade with increasing distance between the human operator and the robot. Therefore, you must design, develop, program, simulate, construct, and finally deploy to the area of

© David Allen Blubaugh, Steven D. Harbour, Benjamin Sears, Michael J. Findler 2022
D. A. Blubaugh et al., *Intelligent Autonomous Drones with Cognitive Deep Learning*,
https://doi.org/10.1007/978-1-4842-6803-2_1

interest (AOI), a fully autonomous AI rover. It must explore these unknown areas without the rover's getting lost either to caves-ins or undetected booby-traps. Because of the possibility of data-link loss, you must incorporate adaptive intelligence within the artificial intelligence (AI) rover. There also exists the genuine possibility of finding the lost treasures of both Pharaohs Khufu and Khafra within these unexplored caverns and wells located at the base of the pyramids (Figure 1-1).

Figure 1-1. *Pyramids and tombs of Giza, Egypt*

Note   The AI rover will need to be as self-reliant as possible to allow it to survive its intended tomb-raiding missions.

Chapter Objectives

The reader will be able to achieve the following by reading this chapter:

- Appreciate why specifications and requirements are important
- Create specifications and requirements for this project

- Understand the basic components of the AI rover

- Realize the importance of selecting the right chassis

- Realize the importance of the Robotic
  Operating System

- Realize the criticality of autopilot

- Realize the importance of Mission Planner software

- Be introduced to the concept of intelligent power
  analysis

Defining Specifications and Requirements

You must first write down what your new AI rover should do to complete
a mission; for example, "Avoid obstacles," "Explore new areas," etc.
Informally, these are the "requirements." Defining the initial requirements
gives us direction in planning development. Formally, the development
of software specifications and requirements (SSR) forces us to first think
about what we want the system to do. We are going to create a land-based
AI rover using cognitive deep learning. The land-based AI rover will be
designed to complete its mission of exploring dangerous and unknown
environments.

SSRs are the key to successful software project development. Let's
now break SSRs into their component pieces. Requirements are what the
system must do (the rover must avoid objects in its path); specifications are
a technical formalization of how the requirements might be met (the rover
will turn left if the object is less than 1 m from the right quadrant). SSRs
evolve as discoveries, corrections, and updates about the system are made.
SSRs take on a life of their own and ultimately transform into a creation
that the reader may not have envisioned at the beginning of development.

The requirements are divided into functional and non-functional requirements. For example, "The rover shall return from a mission upon operator command" is a functional requirement. A functional requirement is a requirement that the human operator usually instigates or supervises. On the other hand, "The AI rover shall return if the battery falls below 50%" is a non-functional requirement. The rover will return without operator input.

Note   The AI rover would function the same if we chose other programming environments (Ada, C/C++, MATLAB, Java, etc.). This is why choosing the programming environment is functional, since we can see the source code within a development environment, or IDE.

Requirements come from many different sources (interviews, observation, forms, developers, etc.) and are not always easily understood. Software designers have used the Universal Modeling Language (UML) to better understand requirements. In particular, a UML use-case diagram helps us understand functional requirements. Further development of the use case will introduce alternative paths to success, pre- and postconditions, and tests. These artifacts in turn allow us to generate specifications. Specifications are the formalization of the requirements. They can be read by any competent programmer as the outline of the function to be written. They are programming language independent.

This implies that without reasonable requirements, we will not have good functional specifications! We will use UML to ensure that all shareholders (technical and non-technical) understand the logic and structure of the system's solutions.

Cognitive Deep Learning Subsystem (Move)

Our rover must search confined areas without GPS or data-link.

Because of this, the system will need to think for itself. To do this, we will use a cognitive deep learning system (Figure 1-2), which is a system that uses a network of smaller deep learning neural networks that cooperate with each other. The cognitive system develops semi-optimal solutions in conditions never before experienced. In other words, the AI rover can autonomously function in new and unexpected environments.

Figure 1-2. *Visualizing a cognitive deep learning system*

The hardware and software components necessary to create the cognitive deep learning system are the Raspberry Pi 4, the Robotic Operating System (ROS), and QGIS. The Raspberry Pi 4 is the hardware brain; the ROS coordinates the sensors the Raspberry Pi 4 needs; and the QGIS is the major consumer of the cognitive deep learning system data (Figure 1-3).

Raspberry Pi 4 Module

QGIS Software   Robotic Operating System

Figure 1-3. *Visualizing the cooperation between the Raspberry Pi 4 module, QGIS, and ROS*

We use the following justifications for our choices of hardware and software components for the cognitive deep learning system:

1. The Raspberry Pi 4 can handle the required extensive computations and commands of a cognitive deep learning controller. It also controls input and output devices to control the speed, direction, trajectory, and navigation. For its size, it has great power and excellent expansion capabilities (USB and general-purpose input-output [GPIO] ports).

2. Robotic Operating System (ROS) controls the simulated and physical AI rover. It allows us to quickly interface our cognitive deep learning system with external sensors and actuators. ROS contains internal libraries such as Simultaneous Localization and Mapping (SLAM) and QGIS for navigation.

3. QGIS displays sensor and location data of the rover on its "map." QGIS is essential for developing paths and waypoints. These waypoints allow the AI rover to revisit areas of interest (AOI). QGIS will communicate with the cognitive deep learning system via ROS.

4. The Pixhawk 4 autopilot will be used due to its low cost and proven reliability.

Basic System Components

Our hardware components start with the GoPiGo robotic rover kit as our chassis. On this kit, we will use the Raspberry Pi 4 as the main processing unit and the Pixhawk 4 as the autopilot. The sensors are a LiDAR sweep sensor and a standard RGB camera.

Our software components start with Ubuntu Linux 18.4 as the host operating system for our project development. The GoPiGo rover uses the Robotic Operating System (ROS). We use Python 3.0 as the programming language. To help in the testing and development of our AI rover, we use the robotic simulators Gazebo and Rviz.

System Rationale (Optional)
System Interfaces

- There will be interfaces between the Raspberry Pi 4 USB ports and multiple devices, such as the Intel neural sticks, Pixhawk 4 interface, and potentially with the ground control station itself.

User Interfaces

- Interfaces will be provided by the ground control station that will be developed within a Python development environment.

Hardware Interfaces

- Four USB ports are located within the Raspberry Pi 4 module. Raspberry 4 can also connect with the Intel neural sticks for higher-level cognitive processing capabilities that will be discussed later.

- Raspberry Pi 4 has a single GPIO port for sensors and actuators.

- Raspberry Pi 4 has a wireless modem for connecting to the internet or the PC or Apple laptop serving as the ground control station.

- Pixhawk 4 has a CAN (controller area network), Ethernet, SPI, I2C, USB, etc.

- Raspberry Pi 4 has a single camera interface.

Software Programming Requirements

- There will be the need to use a programming language (Python) to support low-level devices such as USB, SPI, and I2C.

- Multi-threading: There will be a number of sensors and actuators in the robot. There will be the need to support multi-threading.

- Any Python development environment

- The Raspberry Pi 4 Linux OS firmware

- ROS interfaces with Raspberry Pi 4, Grass GIS applications, sensors, actuators, and autopilot.

- Pixhawk 4 firmware and Mission Planner for Pixhawk 4.

- Python's performance as a programming language. Python has a reasonable ability to work with data being generated by high-bandwidth sensors, such as cameras and LiDAR.

Communication Interfaces

- (Not a completed list. needs to be completed as an exercise.)

- Raspberry Pi 4 module has Ethernet TCP/IP, wireless link (Wi-Fi), CAN, SPI, I2C, etc.

- Pixhawk 4 autopilot has CAN, SPI, I2C, Ethernet, etc.

Memory Constraints

- Raspberry Pi 4 has 1 GB, 2 GB or 4 GB LPDDR4-2400 SDRAM.

- Raspberry Pi 4 can be upgraded with a 500 GB solid-disk drive.

- Pixhawk 4 does have some additional memory to program, but currently is limited to 128 KB (kilobytes).

Design Constraints (Optional)

Operations

- Search, retrieve, locate, and rescue.

- Locate targets and/or threats of interests within unknown environments.

- Explore unknown environments.

- Deliver items or packages.

Site Adaptation Requirements

- (Not at this time. However, this might be subject to change.)

Product Functions

- The ability to communicate between the Raspberry Pi 4, Pixhawk autopilot, and the ground control station; these are the three main communication nodes of an autonomous land AI rover.

- All of the intelligence functions of the AI rover within the Raspberry Pi 4 node will execute the operations stated in "Operations" section just discussed.

User Characteristics

- User has the ability to deploy the vehicle for autonomous terrestrial missions.

Constraints, Assumptions, and Dependencies

- Power constraints of the rover battery

- The terrain that the AI rover might encounter

- As we progress in building this vehicle, additional constraints will be identified.

Other Requirements (Optional)

External Interface Requirements

- The external electrical interface requirements for the autonomous land-based AI rover would be the voltage and current requirements of the interconnecting devices of the Raspberry Pi 4, the Pixhawk autopilot, and the PID (proportional integral differential) controller within the AI rover kit.

- We will find additional external interface requirements within the land-based AI rover itself.

Functional Requirements

- Meet all previously stated requirements.

Performance Requirements

- The real-time characteristics of the AI rover will be revealed to us as we progress through the successive chapters of this book.

- This will also cause us to develop revisions to the SSRs, as well as the subsequent UML diagrams and use cases dependent on the SSR.

Logical Database Requirement

- Not at this time. However, we could supplement the GIS applications found within the land-based AI rover to support database requirements in future versions.

Software System Attributes (Optional)

Reliability

- The reliability of a self-driving and self-exploring land-based AI rover is of the upmost importance. We will see that the software specifications, the UML use-case diagrams, and the AI rover simulators all contribute to a fully functioning system. We will see that as the SSR evolves, reliability can be built into any AI rover system from its initial origins.

Availability

- The autonomous AI rover must then be able to respond to sensor data. It must be able to quickly send information to the AI rover actuators and motors in the form of commands.

- The autonomous AI rover must be able to send data to and receive data from the ground control station.

- The autonomous AI rover must be able to use GIS information efficiently and make decisions based on that spatial data.

Security

- The autonomous AI rover must have limited situational awareness and be able to respond to threats within the environment.

- It must also avoid potential areas in the topology of the exploration area or terrain that would cause the AI rover to become trapped between debris, rocks, or obstacles.

Maintainability

- The AI rover software and hardware must be maintainable with testing, documentation, and upgrades.

- The AI rover software must be documented and maintained within SSRs, UML diagrams, and use-case charts. Any subsequent changes in either the source code of the software or within UML diagrams or use-case charts must be consistent. This means if you make any changes in the source code you must update your UML diagrams to reflect that specific change in software and vice versa.

- The AI rover hardware must also be documented and must have clear schematic diagrams that are compliant with the Specifications. Any changes in the hardware must also be updated within the hardware schematics and or specifications.

Portability

- The use of Python will allow us to export and test the controlling software operating within both the AI rover and the ground control stations across multiple platforms. These platforms include the Raspberry Pi 4, the developing laptop, and possibly even cloud-based internet systems that could be used to test and analyze these same Python routines.

Architecture (Optional)

Functional Partitioning

- There will be partitioning between software components.

- There will also be partitioning between the hardware components.

- We will have to decide as we progress through this book how these functional components will be partitioned.

Functional Description

- The functional hardware description is as follows:

  We have a Raspberry Pi 4 module (with possible expansionary Intel neural sticks) that is connected to an autopilot, whereupon this same autopilot is connected to the PID or control power electronics that power and provide electrical control signals to the AI rover. The autopilot or the wireless connection between the laptop and the Raspberry Pi 4 will allow the land-based AI rover to send and receive information to and from the ground control station.

- The non-functional software description is as follows: We have multiple Python routines operating within the Raspberry Pi 4 that are responsible for the cognitive deep learning routines, the controls for the drive system of the AI rover kit, and the communication software interface with the Pixhawk 4 autopilot. Likewise, there will also be Python routines within the ground control station that will serve as the human operator's remote link with the land AI rover.

Control Description

- The software and hardware must allow the land-based AI rover to avoid obstacles and to complete the mission. Additionally, the human operator can override a decision made by the AI rover if needed.

- The AI rover must interface with sensors, such as gyroscopes, inertial measurement units, accelerometers, and GPS. This allows for the necessary control of the trajectory, velocity, and acceleration of the land AI rover.

- The cognitive deep learning algorithms then will be able to receive data from the sensors. This will allow the rover to efficiently explore an unknown environment within the correct tolerances of trajectory, velocity, and acceleration.

  All of this must be within a closed feedback-control system.

AI Rover Statistical Analysis (Move)

Why use statistical analysis for the development of our AI rover?

Many people who seek to complete a complex project are only concerned with creating a working prototype. Since the Raspberry Pi 4 is limited in its memory resources, we will conduct statistical analysis on the software implementation. The purpose of this analysis is to identify, locate, and eliminate areas within the source code that are inefficient.

We will review techniques that implement optimization for cognitive deep learning networks. Optimization, in computer science, pinpoints the speed-time of programs to increase their efficiency. We will incrementally conduct analysis of the algorithms, implementations, and optimization of the AI rover operating system. We can also utilize these same implementations to find communication issues between the cognitive deep learning nodes, Raspberry Pi 4, Pixhawk 4 autopilot, and the proportional integral derivative (PID) control electronics of the Robot AI rover kit.

We will use these optimization techniques to find communication or control issues. We will also look at the vision-processing routines within the Raspberry Pi 4 to determine if the computer vision algorithms are causing issues with the near real-time processing of the cognitive mission-planning routines or control routines.

Therefore, we will also review the standard questions, such as the following:

- What measurements of the AI rover should I collect?

- What data or command inputs to the AI rover should I test?

- How do I interpret and analyze the data derived from testing the AI rover?

These questions are important and will be answered as we test the rover in a robotic simulation environment. Once we are satisfied with the results of the simulations we can confirm the operating system is fully functional. We will then download the image of our simulation-trained cognitive AI controller into our physical Raspberry Pi AI rover. This is known as creating a "digital twin" for the AI rover.

The experimental algorithmic techniques facilitate better algorithm designs by identifying problem areas. This can improve control and cognitive deep learning routines, such as decision and control, memory hierarchies, and mission analysis.

Bayesian non-parametrics will be used to complete a review of non-parametric regression and classification for cognitive deep learning routines. This determines the uncertainty present within these cognitive deep learning routines and/or models. There will be two examples of this uncertainty testing. The first deals with determining the distance between the AI rover and the edge of a wall. The other identifies the terrain that the AI rover will be operating on. This terrain analysis will include a "go" or "no-go" analysis of the surrounding terrain topology to determine if the terrain is safe for the AI rover to travel. This will also aide in understanding the required mission analysis for the AI rover.

Selecting a Chassis

The centerpiece of our project is the cognitive intelligence engine used for robotic control and mission pathfinding; it creates an extremely adaptable robotic platform. Thus, we could use many different types of robotic chassis. A wheeled chassis will achieve good efficiency, with a simple mechanical implementation.

The additional advantage of utilizing the wheeled chassis is the excellent balance afforded to the robot. This means that all four wheels of the AI rover will be making ground contact, and that the cognitive deep learning engine will not have to adapt itself to the dynamics of an unstable locomotion system. However, the cognitive deep learning engine could make adaptations to problems, such as one of the wheels' being damaged. The AI rover will be required to have a suspension system to maintain constant wheel contact on uneven terrain.

The AI rover will need to focus on problems related to traction, stability, maneuverability, and control. Any off-the-shelf robotic AI rover kit that can be interfaced with the Raspberry Pi 4 and the Pixhawk 4 autopilot can be used with the exercises in this book. The critical concern is whether the wheels provide the required traction and stability for the AI rover to cover the desired terrain. What if the wheels of the AI rover cannot tolerate the velocities, accelerations, quick turns, and trajectories of the AI rover during an autonomous mission? There is a list of commercially available robot kits at the end of this chapter that can be used to complete the exercises contained in each chapter.

Robotic Operating System

The Robotic Operating System (ROS) was designed to be an adaptable platform to support hobbyists, students, and aspiring engineers as they develop robot software. This platform is an assortment of tools, libraries, and development environments to help simplify the difficult task of creating complex and adaptable robotic platforms. These tools include simulation environments for simulating robotic systems such as our AI rover, sensors, physics, time, and dynamics of the AI rover and environment. ROS allows us to take the first steps to a cognitive autonomous controller AI rover system. ROS allows us to incorporate SLAM 3D mapping algorithms and to use the possibilities of geospatial informatics to analyze an environment for targets of interest. All of these capabilities are provided as libraries that can be imported or referenced by our cognitive deep learning program developed within the Anaconda Python development environment.

Pixhawk 4 Autopilot

The Pixhawk 4 autopilot is used for non–decision making control of a vehicle and for simplifying the connections between inputs and outputs. The decision making comes from the operator that evaluates the conditions and status of the vehicle. Autopilots are typically embedded computing systems that control the actuators and PID controls of a vehicle.

The autopilot will be used for error correction and corrective feedback to the internal cognitive deep learning routines within the Raspberry Pi 4. The cognitive intelligence engine and the autopilot will work together to correct *any* errors and return to desired mission parameters. Therefore, we can use two autopilot control structures. The first one would be the position of the AI rover itself, and the second would be rate of speed and/or the trajectory of the AI rover. Therefore, the Pixhawk 4 autopilot will always guarantee that the AI rover will follow the desired position profile. If the cognitive processor decides to move the vehicle at a high speed and turns sharply, this could cause the AI rover to flip over during a mission. The autopilot will prevent this possibility by alerting the cognitive processor to the danger.

The AI rover will receive training from simulations and will receive additional training/reinforcement from operations during the mission. The use of the Pixhawk 4 autopilot adds additional back-up electronics. One of these is the data-link that allows the human operator to take over command of the AI rover. The other advantage of using the Pixhawk 4 autopilot is the Mission Planner software, which allows the human operator to determine the mission status. This allows the AI rover to have remote access to its very own GCS.

AI Rover Mission Analysis

Mission analysis will look at what terrestrial operations the AI rover will be required to execute. This depends on its task. Will the task be a simple "follow-a-path" operation or something more complex?

Therefore, eight rules will be used for each AI rover mission:

1. Define the mission. What are the challenges that are required for the AI rover?

2. Determine tasks to complete the mission. What are the goals that need to be completed by the AI rover?

3. Operational planning. This includes what type of sensors are being placed within the AI rover, power supplies, and possibly even the suspension system of the AI rover.

4. Risk assessment. Are there any associated threats, dangers, or uncertainties within the AI rover mission operations?

5. Communicate risk assessment. Make certain that the engineers, programmers, etc. know what mission the AI rover will perform.

6. Manage and minimize risk (risk management). Make certain the AI rover can effectively avoid all obstacles and dangers.

7. Perform the mission. Execute the mission.

8. Monitor and reassess. Examine how well the cognitive deep learning routines serve as an autonomous controller for the AI rover. Were there any anomalies, failures, or collisions encountered?

AdruPilot Mission Planner Software

The AdruPilot Mission Planner (AMP) is the software that will be the ground control system (GCS), and it can be integrated within a Python development environment. The Mission Planner software gives the AI rover the following capabilities:

- Point-and-click waypoint entry using Google Maps/ Bing/Open street maps. The AI rover goes from waypoint to waypoint and uses its cognitive abilities to sense and avoid obstacles en route between successive waypoints.

- Select mission commands from drop-down menus. Thus, the human operator has ultimate control over the AI rover during mission operations.

- Download mission log files and analyze them. The human will be able to examine how the AI rover performed during a mission (i.e., were there any collisions during the mission?).

- Configure AMP settings for your "airframe," which in this case would be the actual chassis of the AMP-supported four-wheeled AI rover system.

- Interface with a PC flight simulator or programmable development environment to create a full hardware-in-the-loop AI-enabled AI rover simulator. This interface capability will allow the Anaconda Python environment through a GIS library (QGIS, Grass GIS, or even Google Earth) to be interfaced with the Mission Planner software. We can extensively test the cognitive engine before every mission.

- See the output from AMP's serial terminal from the AI rover.

AI Rover Power Analysis

Exploring the caves of Egypt will require endurance from the rover. In the simulation phase of development, a power analysis will be performed to increase the endurance of the rover. This will be improved upon during all subsequent missions to increase the duration and power utilization efficiency continuously. A general principle will be applied to the endurance of the AI rover. When the rover's battery reaches 50%, the vehicle will turn around and return to home. This return to home operation will use the most efficient path as determined by the AI system.

AI Rover Object-Oriented Programming

Object-oriented programming (OOP) has advantages in that development is faster and cheaper, with better software maintainability. Unfortunately, the learning curve is more severe, the software is slower, and it uses more memory. Therefore, we will use UML to help us program the rover.

The OOP paradigm uses the concept of objects and classes. A class can be thought of as a template for objects. These can have their own attributes (characteristics they possess) and methods (actions they execute). All of these objects and classes can easily be modeled and evaluated within the UML format.

List of Components

- Raspberry Pi 4 module

- Intel neural stick USB-drive co-processors

- Robotic AI rover kits

- Any standard laptop with a modern PC processor

- Latest version of Pixhawk 4 autopilot module

- Latest version of Robotic Operating System software

- Latest version of AdruPilot Mission Planner

- StarUML or any other UML development environment

List of Raspberry Pi Rover Kits

- GoPiGo Robotic Rover Platform from Dexter Industries

- Yahboom Raspberry Pi Robot Kit for 4B / 3B+ Project with HD Camera, Programmable Robotic Truck with 4WD, Electronics Education DIY Set for Adult (Raspberry Pi not included)

- Adeept Mars Rover PiCar-B WiFi Smart Robot Car Kit for Raspberry Pi 3 Model B+/B/2B, Speech Recognition, OpenCV Target Tracking, Video Transmission, STEM Educational Robot with PDF Instructions

- 4WD Robot Chassis Kit with 4 TT Motor for Arduino/ Raspberry Pi

We chose the GoPiGo Rover platform for this project. The other kits are valid; however, the GoPiGo Rover platform was our choice.

Acronyms

- SSR (Software Specifications and Requirements)

- GCS (Ground Control Station)

- ROS (Robotic Operating System)

- GIS (Geo-Spatial Information)

- SLAM (Simultaneous Localization and Mapping)

- GPIO (General Purpose Input and Output)

- PID (Proportional Integral Differential)

- LiDAR (Light Intensification Detection and Ranging)

- IMU (Inertial Measurement Unit)

EXTRA CREDIT

Exercise 1.1: What additional changes would you make to the initial SSRs document Version V1.1?

Exercise 1.2: Why are the SSRs so critical for the initial stages of developing deep learning systems for robotics?

Exercise 1.3: Why are statistical methods so critical to finding issues of inefficiency for deep learning systems and robotics?

Exercise 1.4: Can statistical methods be used to identify issues of uncertainty for cognitive deep learning networks? How and why?

Exercise 1.5: What are experimental algorithmics? How would they help us with optimization techniques for the AI rover? Google might be a required reference to answer this question.

CHAPTER 2

AI Rover System Design and Analysis

We aim to develop an autonomous AI rover that can execute an exploratory mission in the Egyptian catacombs. Artificial intelligence (AI) systems such as the AI rover are complex and challenging to understand. Using the informal background story from Chapter 1, we will "draw pictures" to better understand our complex problem. We will use Universal Modeling Language (UML) diagrams to understand the structure and behavior of the system. Models are simplifications of complex ideas. For example, Newton's Second Law of Motion tries to explain the physics of momentum. It is an excellent approximation, but it is not 100% accurate. It leaves out gravity, friction, and so on. But it *is* helpful.

> *"All models are wrong, but some are useful."*

> —George Box

So, to help us understand the complex rover system, we will model the stuff we don't understand well using UML diagrams, which are helpful in modeling structure and behavior. The best thing about UML is that it can help us understand a fuzzy, poorly defined problem, taking us from a very informal description of the problem to its formal requirements and

© David Allen Blubaugh, Steven D. Harbour, Benjamin Sears, Michael J. Findler 2022
D. A. Blubaugh et al., *Intelligent Autonomous Drones with Cognitive Deep Learning*,
https://doi.org/10.1007/978-1-4842-6803-2_2

specifications. You do not need fancy software to draw UML diagrams, as you can draw them by hand. One of the significant advantages of software developed with UML is that some programs will generate simple skeleton codes of the structure defined by class diagrams.

First, we model the problem at a very high level. We do not have to understand everything up front, just enough to "scope" the problem. We stop if we understand our problem well enough to create a complete solution. If we still cannot understand how to solve our problem, we expand the parts of our model we do not understand completely to understand a little more of the solution. We can start developing a solution, and if we get stuck, we can back off and reexamine this part of the problem by modeling it in more detail (i.e., adding a friction component to Newton's equation.)

Although UML diagrams look very similar to caveman art, they are helpful for high-level modeling or understanding a complex system. So, let us start making "useful" models.

Chapter Objectives

The reader will be able to achieve the following by reading this chapter:

- Review the software specifications and requirements for the AI rover.

- Use a Universal Modeling Language (UML) framework for object-oriented software development for the AI rover.

- Create UML use cases to satisfy requirements.

- Understand the difference between functional and non-functional requirements.

- Identify functional and non-functional requirements for the AI rover.

- Learn how to handle complexity within the AI rover.

- Understand the five levels of software and hardware design.

- Review what software tools will be required for this chapter.

- Understand the AI rover's static and dynamic features.

Placing the Problem in Context

Understanding what belongs to the system and what does not belong to the system is our first step. This defines the context of the system. To accomplish this, we first need to define the "black box" that represents our rover. A black box system means that we know it has inputs and outputs, but we do not know how the inputs give us the outputs (Figure 2-1.)

Figure 2-1. *Black box system*

Defining our system should be done in simple English, not engineering technobabble. Our first attempt at writing a goal is "an intelligent rover that can explore unknown environments." This is simple and straightforward, but is it complete? At this time, we will consider it "complete enough," but we may expand it later when we understand the problem better.

As written, it gives us a description of a system that states what is in the system (explore mission, intelligence, etc.) and what is not in the system (search and rescue missions, robot killers, etc.). Although a context diagram is not UML, we will create a simple one—first we start with our system description and treat it as a black box.

Next, we try to figure out who or what will interact with our black box. The things that interact with the system are called actors. In our book, actors are drawn as ovals.

In our exploration problem, we start with one actor. The actor "operates" the rover like a remote-control car by "starting the rover," "navigating the rover," and "ending the mission the rover is on" (Figure 2-2).

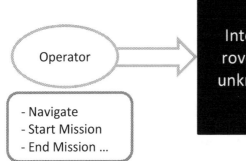

Figure 2-2. *Adding the actors*

Next, we attempt to define what the interactions will be. We do not care how the interactions will be implemented, just that they are listed. The supervisor does not directly control the rover like a remote-control car, but rather the supervisor plans the mission and tells the rover to "start the mission," "stop," "return," etc. (Figure 2-3). The operator may direct the rover to "stop," "turn left," etc. Since the rover is autonomous, it must observe the environment and supply input to itself to "change direction," "stop," "return," etc. Notice that each interaction is of the form *verb-noun* or *verb* .

The last step in our context diagram is to figure out what limitations are on our system. These are known as constraints and may be based on hardware, software, or business reasons. For instance, we will use a Raspberry Pi 4—this means we have a quad-core 1.5 GHz Arm Cortex-A72–based processor. It will not run faster. It can have up to 4 GB RAM and no more. The Pixhawk chip will be our autopilot system, thus limiting us to its functionality. We will use less than $500 in purchasing the entire system. These are all constraints that are necessary to document at the beginning so that we limit our search for a solution to within the boundaries of the constraints (Figure 2-3).

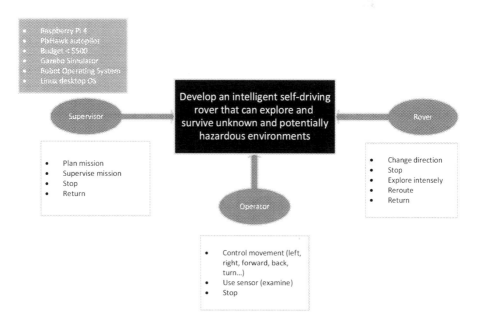

Figure 2-3. *Context diagram with constraints*

At this point, we have enough to begin to understand the problem. We don't have enough information to complete the solution, but we have a glimmer of the complexity. Obviously, the project is bigger than this diagram suggests, but we can modify the diagram as we more fully understand the problem. The context diagram also starts to help us to understand the functional and non-functional requirements. The constraints are examples of non-functional requirements, and the lists below each actor are examples of functional requirements.

Developing the First Static UML Diagrams for AI Rover

We will continue our discussion of developing a rudimentary adaptable navigational system for the AI rover. We should state that what is meant by the word "adaptable" in this sense is that the robot is able to stop and avoid collisions with obstacles encountered within its driving path. We will see in future chapters of this manuscript that we will develop sense-and-avoid capabilities that will allow the rover to develop alternative paths to help it to circumnavigate obstacles located within its path. The future development of the AI rover's ability to determine alternative driving paths will be critical for it to avoid obstacles encountered during its decent into and exploration of the well, caverns, tombs, and catacombs located at the base of the Great Pyramids of Egypt.

The purpose of doing static modeling for each of the components of the AI rover is to design the interfaces between the hardware portions of the AI rover, the software components within the AI rover, and the environment (simulated or physical). This is done to create the static structure of the AI rover under development so as to develop the class diagrams for (in this case) the beginnings of the adaptive navigation component of the AI rover. This is especially important for developing the Python routines that are needed for the near-real-time nature of an AI rover. Therefore, the system-context class diagram can be developed by static modeling the system classes that are connected to the different software and hardware components of the AI rover. For example, please refer to the class diagram in Figure 2-4.

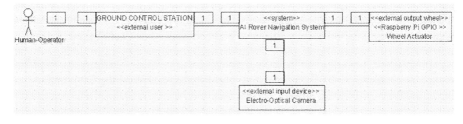

Figure 2-4. *Class diagram for camera navigation*

Now let's review the contents found within Figure 2-4. We have now developed the context class diagram for the first use case that was developed in Figure 2-3. The context class diagram reveals how the system needs to interface with external devices or actors. The navigation system for the AI rover accepts a command from the human operator through the ground control station. This command then tells the robot to what destination it needs to travel within the grid-map. In our case, we have selected an electro-optical camera to be interfaced with the AI rover navigation system. There are other sensors that could be employed with this interface, such as LiDAR, ultrasonic, and proximity sensors. The critical need for the context class diagram corresponds to the interactions between the AI rover's navigational system, the connected camera sensor, the output wheel actuator device via the Raspberry Pi's GPIO, and the external interaction with the ground control station and the human operator. We could also easily add a timer to the system to periodically send update commands to the wheel actuator, to allow the wheel actuator to be more reactive to direction updates.

The primary purpose of developing static class diagrams is to allow us to develop the foundations for creating the dynamic modeling for the adaptive navigational component for the AI rover. Another primary purpose of developing that static model is to resolve the problem of the navigational component into objects within the system. A critical part of this process is to identify the external interfaces of the navigational component. These interface objects would be the wheel actuator interface, the ground control station interface, and the sensor interface. Now we will need to consider how to develop the control objects that are necessary to allow the AI rover to operate correctly. These control objects are critically needed to provide coordination and collaboration between the objects defined within both the use case and the static modeling of the navigation component. Examples of control objects would be the coordinator, state-dependent control, mission timer, and camera control or camera vision signal processing. As such, there will also be storage memory objects, which store memory that needs to be contained in order for the navigational controls to be operational; they would be the navigational map (could use QGIS), driving pathway determined, destination, and current map position (possibly data from GPS). The use of the current map position and the mission timer could allow us to develop a prediction as to the mission duration and power-rate consumption characteristics of the AI rover's moving from its current position to a targeted position. We could also use the mission timer to determine if the AI rover has fallen into a position that it cannot escape from and whether the rover might need to turn the electrical power off and conserve power as a result of the unsatisfactory period of time that the AI rover has taken to reach its intended target location. This might be true in the event that the rover has either flipped over to one of its sides or has become trapped and can no longer proceed to the final destination point. Events such as this need to be handled by our mission timer. We could also designate our mission timer and the navigational timer as being one and the same. The mission or navigational control timer will be controlled by the navigational control object. We also need to point out to the reader that the navigational control object itself will ultimately be controlled by the cognitive deep learning intelligence engine object that we are now in the process of creating.

As we develop the object structure for our navigational system, we can see that we can add localizer and path-planner algorithms to the internal structure of the navigational system. By taking advantage of the object-oriented programming paradigm, we can test different types of localizer and path-finding algorithms to statistically analyze, test, and determine which one of those groups of algorithms is the most computationally efficient. For example, we could determine through statistical analysis whether the A* algorithm would be ideal within the role of exploring the Egyptian catacombs as the main path-finding algorithm. As such, we can review the class structure diagram of the navigational system shown in Figure 2-5.

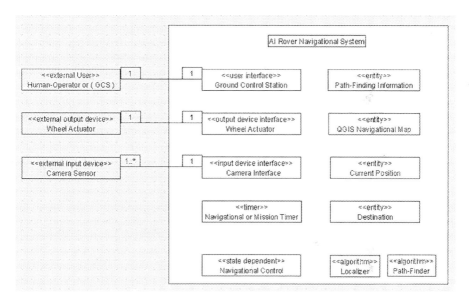

Figure 2-5. *AI rover system class-association relationship*

Developing the First Dynamic UML Diagrams for AI Rover

The purpose of dynamic modeling is to capture and understand the dynamic nature of the AI rover system so we can understand its subsequent components. It is important to note that dynamic diagrams make use of the interactions that were identified during the initial use-case and static-diagramming process for object structuring of the AI rover navigational system, as identified in Figure 2-5.

By now, it should be evident how an object-oriented approach to design can allow us to develop systems with scaling complexity. We are now able to visualize the first robotic structures in software and hardware which are starting to take form. We can now start to develop the modular control structure, which is composed of, at first, the deep learning networks, and later the cognitive architecture that will be constructed on top of these deep learning networks. This modular control structure will also allow us to create specialized components that pertain to the interfaced components, such as LiDAR and camera sensors. Each of these software components will act as a brick in the final creation of an AI rover.

The Robotic Operating System (ROS) that we will be reviewing in Chapter 3 will also allow us to develop a general-purpose robotic control software package that will be based on components both in simulation with AirSim and Gazebo and within the real-world system. The already present nature of modularity and flexibility that is available within recent variants of ROS will allow us to experiment with and develop complex robot systems with cognitive engines built in. The use of ROS will allow us to also integrate components for kinematics, dynamics, control, decision making, limited situational awareness, and software–hardware interfaces. Each these items will be either a class or an object that can be added or removed within the open-source development environment for ROS.

One last step we need to review before we go into developing the actual dynamic model is the business process for the robotic navigational task that was discussed in the last section on static modeling for the AI rover. This task that is being controlled at first by our deep learning routines and then finally our cognitive engine that will be built on top of it will most definitely need to modeled within an object-oriented framework because of the number of interacting and controlling components. The one problem that needs to be addressed is self-localization as one of the main goals for navigation. The information sent to the navigation system is provided as the direction and distance from the goal to target. The AI robot must also deal with errors caused by sensors and environmental uncertainty. Therefore, we will go over the business model diagram, which represents how the mobile robot cognitive deep learning navigation system must operate. This system was developed with a hierarchical paradigm in mind. The AI rover, as well as the decision-making cognitive deep learning engine, acts only after planning the driving pathway by processing sensor readings. The cognitive deep learning system allows the robot to enable a navigational task with continuous sensing in a static environment where there are no objects moving, as within the Egyptian catacombs. Once the navigational task is being executed, the AI rover will complete a new cycle of information from all available sensors and then finish with a single move command (Forward, Backward, Left, or Right) produced by the cognitive deep learning engine. Eventually, in the later chapters of this book, the amount of distance covered with each Forward or Backward movement command and the degrees for the angle of attack with each Left or Right command will ultimately be produced by the cognitive deep learning engine itself.

Note   Please be aware that we will be closely reviewing and describing the subsections of the business logic that will describe the navigational and cognitive deep learning processes needed for the AI rover to explore the catacombs and external structures of the Egyptian pyramids. This business logic is critical in helping us to create the UML class diagrams. This portion of the text needs to be closely read and examined for comprehension.

Therefore, the first section of this process has to start at the beginning of the business logic diagram itself. When the adaptive navigational task begins, the first process, in the form of a box, will obtain sensor readings from LiDAR, ultrasonic, proximity sensors, and Raspberry Pi 4 cameras. The first process (processes are indicated in the diagram as boxes) executed is the sensor readings. This information is then sent to update the resource database of Sensor Data. All databases are represented by cylinders in business logic diagrams. Sensor Data is a list of all sensor spatial coordinates retrieved by the Sensors process, and the data will usually utilize the cartesian coordinate system. These coordinates are usually the ending points of the vectors starting at the robot and ending in the detected obstacles within the Egyptian catacombs. What this means is that the coordinates indicate the position that the robot was in when it took the sensor sample and the corresponding calculated distance and coordinates of the sensor data point of the object that was detected by the same sensor. Figure 2-6 represents the UML business logic diagram of this very first stage of the adaptive navigational task for the AI rover.

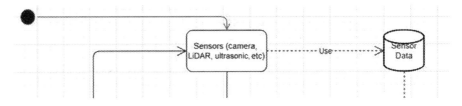

Figure 2-6. *Business logic diagram for sensors and sensor database*

Note   Please note that every process and database for the AI rover's adaptive navigational task will be fully described conceptually and logically within their correct and corresponding chapters in this book. Remember this AI rover needs to survive.

Once the Sensors process has completed, there will be another process to be executed. This next process is called the SLAM or Vision Processing Analysis process. The SLAM, or Simultaneous Localization and Mapping, is a statistical algorithm for processing either a two-dimensional or a three-dimensional map and thus determining the location of the robot within the ever-expanding map. We will also talk about how the map process of SLAM might be limited due to memory allocation issues for such a problem. For the purpose of this version of the AI rover, we will be concentrating on a two-dimensional representation for SLAM in order to determine the location of the robot and what course adjustments might need to be made as the AI rover is carrying out its mission in the catacombs. The local map is based on the estimate of its position, and this same local map is then integrated within the global map to help the AI rover navigate toward its goal location. SLAM utilizes multiple estimated values that can then integrate information from several sensor sources. SLAM then determines the position of the AI rover and updates it to the Sensor Data database. Another algorithm that can be used within SLAM or image processing is to use monocular or binocular cameras to determine the position of an object and to execute sense-and-avoid maneuvers for the AI rover as it is going from one point to the next.

Once the Position and the Sensor Data databases have been updated, the next step is to execute the Occupancy Grid process. This process transforms the sensor data within a tessellated map of the probability of obstacles' being within a certain point or region of the grid. It then uploads the local map, which is why the local map must be current and centralized constantly at the AI rover's location. Once this is done the local map is then incorporated into the global map, which could also use the GIS software package QGIS as the database. Figure 2-7 shows the UML business logic diagram that describes this process within the AI rover's adaptive navigational task.

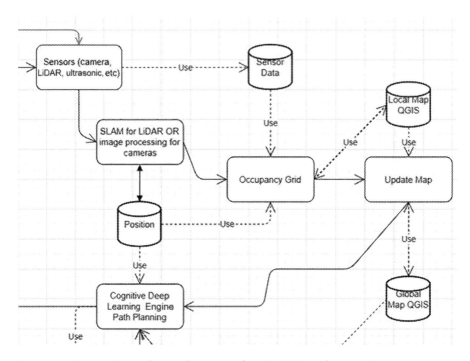

Figure 2-7. *Business logic diagram for SLAM and cognitive processing*

The final process involves cognitive deep learning, which directly controls the path-planning process, one of the primary responsibilities and characteristics of the cognitive intelligence engine. The cognitive engine is thus provided with the most current version of the global map database, as well as the required goals of the navigational task and the estimated location of the AI rover. The driving path solution is then generated by the cognitive deep learning engine and is then followed by the rover itself. The path is essentially cartesian coordinates interpolated by straight lines for the rover to traverse. Therefore, each line is a specific driving action for the AI rover's wheel actuators. Also, the pathway is determined by the physical dimensions of the rover, as it determines whether the rover has clearance with respect to areas around the rover itself. The cognitive engine also regularly updates the path segment. The path segment is what actually and directly gives the commands to the wheel actuators as to how to follow the pathway driving solution. If all of the goals have NOT been completed, then the entire cycle begins again. That is the reason for the diamond-shaped decision block in the UML diagram in Figure 2-8.

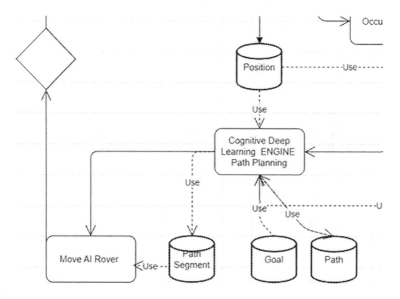

Figure 2-8. *Business logic diagram for path segmentation and move commands*

Note   Now the entire adaptive navigation task that is being controlled by the cognitive engine can be appreciated within the following UML business logic diagram (Figure 2-9).

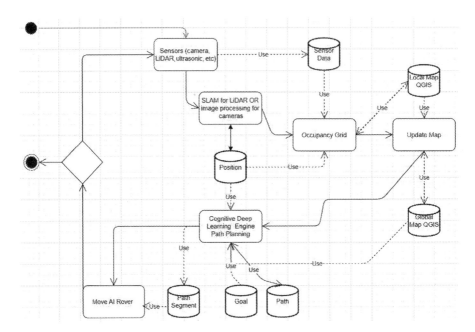

Figure 2-9. *Entire business logic for the AI rover system*

Developing the First Dynamic UML Class Diagrams

Developing the adaptive navigational requirements means that the goals stated by the human operator must be obeyed and followed, including by the cognitive deep learning engine that is operating within the AI rover. Therefore, we will set the goals or the tasks as the primary super-class that takes precedence over the entire structure of the software system that expresses itself as the software architecture of the AI rover. The Goals super-class is the class that is responsible for the goals that are to be followed by the AI rover. Therefore, the GOALS super-class will utilize both the AI_Rover class and the Cognitive_Deep_Learning_Planning class. The main functions or algorithms of the GOALS super-class are to establish the goals and to check which goals have been completed. Another method could represent the progress or percentage completed of an AI rover exploring a particular Egyptian tomb or catacomb. In addition, the use of GOALS as a super-class helps to create a rudimentary subsumption architecture, where the goals of the system are obtained and that tells the AI rover to go to each goal point.

Therefore, we will now create the UML class diagram. We will add each level of classes one at a time to allow the reader to have a complete understanding of the UML class architecture of the AI rover. This is essentially the entire program structure for the AI rover, with all the correct interconnections. Therefore, the first class that we will be working on is the GOALS class, which states all of the goals that need to be completed by the AI rover. Figure 2-10 shows the UML class diagram for the GOALS class.

⊞   GOALS

Figure 2-10. *AI rover Goals class diagram*

There are two main sub-classes that are connected to the GOALS super-class. The first connected sub-class is the AI_Rover class. This sub-class is the one responsible for describing the type of AI rover that will be used within the cognitive deep learning system. The AI_Rover class contains a composition of at least one wheel actuator and one sensor. The Wheel_Actuator class is the one responsible for interfacing with and sending commands to the motor controllers that drive the AI rover itself to the next waypoint or pathway determined by the cognitive deep learning controller. The Wheel_Actuator class then has additional sub-classes, such as the Heading and Steering_Controls classes. The Heading class determines the consistent velocity of the AI rover traveling within the pathway determined by the cognitive deep learning controller. The Steering_Controls class is responsible for assigning the correct steering or (wheel-turning) velocity of the wheel actuators turning the AI rover. The Heading and Steering_Controls classes will usually be interfaced with the

PID (partial integral derivative controllers of the rover platform that will be determined. Usually, these PID controllers will be manifested as hardware closed-loop controller components outside of the Raspberry Pi 4 module.

The next sub-class that is connected with the GOALS super-class is the actual Cognitive_Deep_Learning_Planning class. This class is responsible for the correct decision-making analysis, limited situational awareness, and sense-and-avoid properties of the AI rover conducting its missions within the Egyptian catacombs. The prime directive of the Cognitive_Deep_Learning_Planning class is to develop the correct path-planning solutions to be used for the critical navigation of the AI rover. This navigation also needs to be adaptable, so that the AI rover can find an alternative path in the event it encounters an impasse. Therefore, the Cognitive_Deep_Learning_Planning class must utilize the MAP class, which is a super-class of both the local and global maps. The main concern for the Cognitive_Deep_Learning_Planning class is to achieve the goals stated within the GOALS super-class. The Cognitive_Deep_Learning_Planning class generates both the whole path and the actual path segment to be sent to the AI rover's wheel actuators. The Cognitive_Deep_Learning_Planning class will need access to methods to retrieve and send information regarding the robot's next path segment, which is planned with respect to the AI rover's current position. Some of the methods of this class are Set_Goal, Determine_Global_Path, and Get_Path_Segment. These methods will be used by the cognitive deep learning controller in the event that the robot needs to find an alternative driving path segment to follow, with the time allotted before the AI rover's battery becomes decapacitated. Figure 2-11 shows the UML class diagram of the classes that we have reviewed and discussed so far.

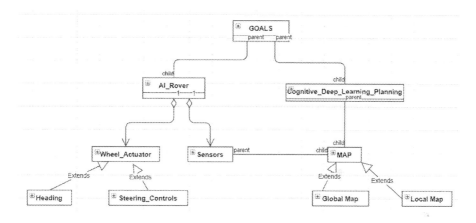

Figure 2-11. _First stage of the UML dynamic class diagram for AI rover_

The Sensors class is the class that many other sensors can be developed from within the architecture of the AI rover. We have derived the sensors of force, proximity, tracker, range, position, velocity, and acceleration. We can derive other sensors from the Sensors class, such as LiDAR, Radar, and monocular and binocular cameras. All of these sensors can be developed based on the requirements of the shareholders and developers. We will need to have discrete increments of time, since we are operating within a digital platform such as the Raspberry Pi. This would be an example of time_increment within the discrete time steps the AI rover's architecture must complete. The other methods would be START, STOP, SetUpdateTime, GetUpdateTime, S_Variance, G_Variance, and READ. The READ function returns information derived from the sensors. The position, velocity, and acceleration sensors and the classes that develop them obtain information from their respective sensors. For example, the position class derives its information from an encoder, which gives faulty values over time. However, the use of the map system derived from the SLAM and vision processing sensors, such as cameras, can allow for these erroneous position values to be corrected by the localization methods derived by the MAP class. There is one last class in the second stage that we need to discuss. The Range class represents the range sensor within the AI rover. This range sensor could in reality be an algorithm that processes information from a monocular camera (a camera with only one lens) and uses a form of triangularization to determine the distance to targets based on estimating the distance from the edges of objects to the position of the AI rover. However, typically these sensors are either binocular cameras, laser range finders, or ultrasonic sensors. These sensors thus are responsible for estimating the distance between the surfaces of objects and the position of the AI rover. The range sensor also has a critical role to play in the creation of the MAP and SLAM processing classes found in the first stage of the AI rover's architecture. It is important to consider eliminating the blind spots that are found within the SLAM processing of the passageways of the Egyptian pyramid complex and subterranean catacombs and tombs. Let's discuss the sensors that will help in that regard.

The first sensor is the LIDAR laser sensor and/or class. This sensor can detect the presence or absence of an object in front of the AI rover. This determines the azimuth angle (direction) in which the object is located with respect to the AI rover's position.

The tracker sensor and class is responsible for the visual tracking of a target. However, this functionality might be incorporated directly within the cognitive deep learning engine. Therefore, this functionality might instead be incorporated within the Cognitive_Deep_Learning_Planning class. However, we could make this process a secondary process and not have it be developed within the cognitive system.

The Stereo_Vision class also uses the binocular cameras, in order to determine the range between a target and the location of the AI rover. This will require information from the camera as well as a range-finding sensor. The range-finding sensor could be either a proximity sensor or a laser range-finding sensor. Therefore, we will now reveal the final piece of the main architecture view of the cognitive deep learning system, which has been combined with the rest of the controlling structure of the AI rover. The UML class diagram is as shown in Figure 2-12.

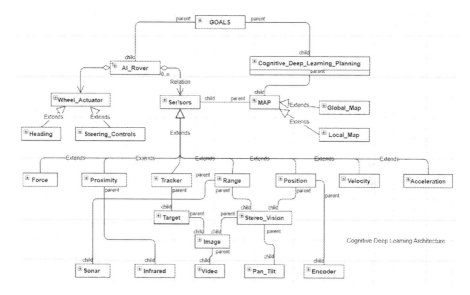

Figure 2-12. *Second stage for the UML dynamic class diagram for the AI rover*

Developing the First Dynamic UML Sequence Diagrams

One of the final dynamic UML diagrams that we will be reviewing is the sequence diagram. The following sequence diagrams will allow us to closely examine the interactions between the objects that allow the AI rover to autonomously operate within the Egyptian catacombs or any other unknown environment. These diagrams also reveal the time events of when messages were sent to different objects within the software architecture itself. They also show which methods are called by their respective objects within the AI rover. If we review the global map, we see that it allows for both static and dynamic changes within the environment encountered within the Egyptian catacombs. We can also see how SLAM—or the simultaneous localization and mapping—changes with new sensor data obtained from LiDAR, ultrasonic, or even monocular and/or binocular cameras allows the local map (immediate area around the AI rover itself) to update itself and the global map. The updating process of the global map allows for correction of the estimated position of the AI rover within the Egyptian catacombs. This is especially true when the unknown environment within the catacombs for some reason has unexpectedly and drastically changed. An example of a drastic change within the catacombs could be that there has been a catastrophic collapse or cave-in and there is structural destruction that has denied access to the AI rover's driving pathway solution. This would mean that the isGlobalMapChanged method will indicate that the known global map of the Egyptian catacombs has been completely disturbed, and that there is a need for the cognitive deep learning engine to derive another or updated driving pathway solution for the AI rover to execute in order to complete its mission objectives.

We will now review the following dynamic UML sequence diagram (Figure 2-13) for the case where there have been no dramatic changes within the ancient Egyptian catacomb environment being explored by the AI rover and thus no global map updates. The following will be a chronological list of events:

1. We can first see that the USER class has set and defined goals and initiated the Start process.

2. Once that has been completed, the GOAL class sends a start mechanism to the AI_Rover class, which will power up the PID electronics. We could also further develop this sequence diagram to include an additional diagram to execute a diagnostic check for the PID electronics to determine if the rover kit itself is correctly operational. Please consider that a possible exercise for this chapter.

3. The AI_Rover class then proceeds to start the Position class methods, which means gathering information from the GPS of the PixHawk autopilot and/or from the global map for the location estimate.

4. The AI_Rover class then starts the range class methods, which usually means receiving information from the global map.

5. The GOALS class then translates the required goals and tasks to be completed by the cognitive deep learning engine, which is acting as the controller of the AI rover, to then derive a driving pathway solution.

6. The cognitive deep learning controller then makes a request to determine the physical and geospatial location of the AI rover within either the entranceway of the Egyptian catacombs or within the catacombs themselves. This request is sent from the cognitive deep learning controller directly to the AI_Rover class itself.

7. The AI_Rover class then sends a READ command to the Position class that is handling information between the PixHawk autopilot, the PID electronics of the rover kit, and the Raspberry Pi 4 module.

8. The AI_Rover then handles the return PositionValue coming from the Position class.

9. The PositionValue data is then passed from the AI_Rover class back to the requesting Cognitive_ Deep_Learning_Planning class.

10. The cognitive deep learning engine then requests the GetMap method of the Global_Map class to retrieve information from the current global map, with the AI rover at the center of the map.

11. The MAP_DETAIL data and request is then sent back from the Global_Map class to the Cognitive_Deep_Learning_Planning class.

12. Then the cognitive deep learning controller will determine the best driving pathway for the AI rover to execute given the current information obtained from the global map.

13. The FindPathWays method is then executed within the cognitive deep learning engine to calculate and determine the best pathway with the current information.

14. There is then a PathwayFound message sent from the Cognitive_Deep_Learning_Planning class back to the GOAL class to inform it that a driving solution has been found.

15. A HasMAPChanged message is then sent from the Cognitive_Deep_Learning_Planning class to the Global_Map class.

16. If the Global_Map class returns a message of NO back to the Cognitive_Deep_Learning_Planning class then the Cognitive_Deep_Learning_Planning class sends a ReadPosition request message to the AI_Rover class, which then sends the message of READ to the Position class.

17. The PositionValue message containing position data is sent from the Position class back to AI_Rover and then from there back to the Cognitive_Deep_Learning_Planning class.

18. A PathSegmentDerived message is then sent to the GOALS class.

19. The GOALS class then sends a MOVE command with the PathSegmentDerived solution as an argument to the AI_Rover class, which is then sent as a message to the Wheel_Actuator class. The rover by this time should start moving to drive the path segments determined by the Cognitive_Deep_Learning_Planning class, acting as an adaptive navigator within the software hierarchy of the AI rover.

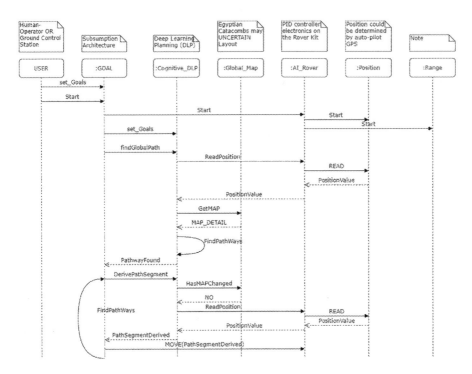

Figure 2-13. *UML sequence diagram for cognitive navigation and controls*

The following UML sequence diagram (Figure 2-14) will describe the process of updating both the local and the global maps based on either SLAM or image-processing data updates. This happens in the cases where the environment that the AI rover is exploring is either static or is changing and dynamic.

1. The AI_Rover class sends a start message to the Position and Range classes.

2. The local map then sends a READ and retrieves a RANGE message to and from the Range class.

3. The local map then sends a READ and retrieves a POSITION message to and from the Position class.

4. The local map then updates itself.

5. Then the local map updates the global map.

6. The SLAM_VIDEO_Proc method is then executed within the local class, and then it again updates the local map.

7. The AI rover is then centered within the local map. This process is continually cycled ad infinitum.

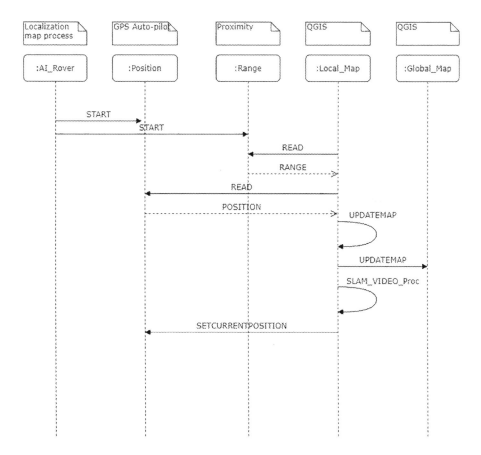

Figure 2-14. *UML sequence diagram for SLAM and video processing*

The following UML sequence diagram (Figure 2-15) is one in which there would be a drastic and detectable change with the global map or even local map within the Egyptian catacomb environment. An example of this would be if a floor or a passageway within the Egyptian catacombs were to unexpectedly collapse. As such, is it is critically important for the AI rover to be adaptable enough to calculate a new driving path solution. The cognitive deep learning engine should have some degree of limited situational awareness to determine if continuing the mission is advisable or is dangerous. This process is described within the following steps and procedures:

1. The UPDATEMAP message is sent from the local map to the global map, whereupon the global map is updated.

2. The GOALS class then sends a DERIVEPATHSEGMENT request message.

3. The Cognitive_Deep_Learning_Planning class then sends an isGlobalMapChanged request message to the Global_Map class.

4. If the Global_Map class returns a YES response message, then the Cognitive_Deep_Learning_Planning class sends a READPOSITION command to the AI_Rover class.

5. The POSITION data message is then returned to the Cognitive_Deep_Learning_Planning class, which then requests a GetMAP message from the Global_Map class, which is then returned as a MAP_DETAIL data message.

6. The Cognitive_Deep_Learning_Planning class
    then finds alternative pathways and sends that
    information to the GOALS class.

7. The GOALS class then sends an alternative driving
    solution with MOVE(PathSegmentDerived) to the
    AI_Rover class, where the rover's Wheel_Actuator
    class should be making the corrective heading and
    steering adjustments.

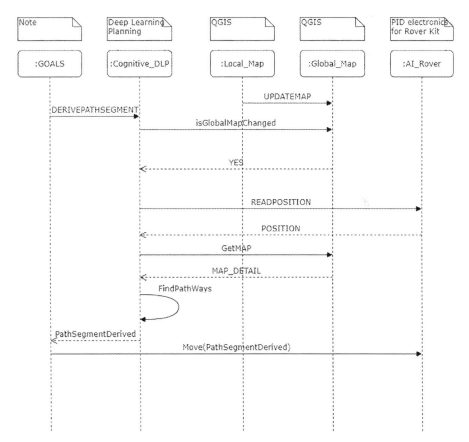

Figure 2-15. *UML sequence diagram for drastic change to*
global map

Summary

We should state that we have now reviewed and become familiar with multiple software and system design analysis approaches. One approach was in regards to developing UML use-case diagrams that can help us to determine both the functional and non-functional requirements for developing this fairly sophisticated self-driving, adaptable, and intelligent rover system. The next UML diagrams that we reviewed were the business logic and class diagrams, which showed how the modular classes relate to each other and how they might send information to each other. For example, consider how the `Cognitive_Deep_Learning_Planning` class controller serves as an adaptable navigator for the AI rover itself. The use of these UML diagrams allows us to quickly specify and review some very difficult complexities involved in the creation of the AI rover self-driving platform.

We have also revealed how an adaptable intelligent robot system can be modeled with UML object-oriented concepts. The position of the robot is at first corrected by SLAM, and the planning is conducted by the `Cognitive_Deep_Learning_Planning` class controller within a grid system. We have considered both static and dynamic environments, and we have developed a software hierarchy that might allow for what would otherwise be a very computationally expensive approach to navigating dynamic environments. The objects were shown within these UML diagrams, as well as how they interact with each other. All of these classes were controlled by the `GOALS` class, which acted as the commander or pilot, and the `Cognitive_Deep_Learning_Planning` class controller acted as a kind of adaptive navigator for the AI rover exploring the netherworlds of the Egyptian catacomb complexes.

We should also state that the use of these UML diagrams can allow us to develop the implementation of the controlling software system for the AI rover, in that these same UML diagrams can be auto-generated into Python source code. There are multiple development environments that can allow for this to happen, such as Modelio UML, Papyrus IBM Eclipse, Rational IBM software, Microsoft Visual Studios, and many others. By using the UML diagrams, we can help manage the complexity that might arise with such an advanced robotic rover platform. If there are any specification or requirements changes, we can manage those as would be expected within any software development project.

EXTRA CREDIT

1. In reviewing the business logic diagram, where would one need to include an autopilot?

2. In reviewing the class diagrams, again, where would one need to include an autopilot?

3. Why would UML diagrams help us in understanding class interactions, data, and message exchanges?

4. What is the concept behind the "Unexpected Query Problem" for the AI rover? We will discuss this concept in later chapters, but please use resources on the internet to find and derive your best answer to this question.

CHAPTER 3

Installing Linux and Development Tools

Before we start developing our rover, we have to install the right tools. Just like a carpenter needs a hammer to nail and a painter needs a paintbrush to paint, we need tools to support our project. Our rover will use the Robot Operating System (ROS), and ROS runs on Linux. Linux gives us a simple text editor and the Python programming language. We also need a simulator to test whether our rover works; thus, we will install ROS-aware Gazebo and Rviz. Finally, we assume most of our audience will be using the Windows operating system, so we will install all of these programs on a virtual machine using VirtualBox. In summary, we will review the following:

- Installing Oracle's VirtualBox in Microsoft Windows

- Installing Linux Ubuntu 20.04.4 in VirtualBox

- Acquiring the ROS environment variable key

- Installing ROS

- Starting ROS for the very first time

- Learning important Linux ROS shell commands

- Installing Anaconda

© David Allen Blubaugh, Steven D. Harbour, Benjamin Sears, Michael J. Findler 2022
D. A. Blubaugh et al., *Intelligent Autonomous Drones with Cognitive Deep Learning*,
https://doi.org/10.1007/978-1-4842-6803-2_3

- Learning Ubuntu Linux commands necessary to develop rover prototypes

- What are ROS launch files?

- Running Gazebo and Rviz to test our system

- Summary, exercises, and hints

Before We Begin

We must consider the following preconditions:

1. If you are planning on using the Ubuntu Linux operating system as the host operating system, then there is no need to install VirtualBox.

2. Different versions of ROS target specific versions of Ubuntu Linux. We will use Noetic ROS, which specifically requires Ubuntu Linux 20.04.4. Make sure you match the ROS version to the correct Linux version. This is important during "Installing VirtualBox ..." and "Installing Ubuntu Linux ..."

3. The ORDER of installation is important: VirtualBox, Ubuntu, Anaconda, ROS. Any other order may cause problems.

4. Ubuntu Linux 20.04.x is also known as Focal Fossa. We will call it Ubuntu. We will use Noetic ROS, but we will call it ROS.

5. Although all shell text commands will be for Noetic ROS, they are compatible with earlier versions of ROS, such as Kinetic and Indigo.

Installing the VirtualBox Software

The purpose of the Oracle VirtualBox is to create virtual machines to develop and run programs in an operating system other than your regular system. The virtual machine is an emulation of the target operating system. We want the Ubuntu Linux 20.04.04 OS emulated on our Microsoft Windows system. (There are versions of VirtualBox for Mac OS and Linux.) After installing VirtualBox, we will install an Ubuntu Linux 20.04.04 virtual machine. This virtual Ubuntu Linux 20.04.4 OS will serve as our ROS development environment. This virtual machine is essentially a program that executes an entire operating system, such as Ubuntu Linux 20.04.4 OS, as an application. The physical computer, running Microsoft Windows, serves as the *host*, and Ubuntu Linux 20.04.4 serves as the *guest*.

The Ubuntu Linux guest operating system in VirtualBox will be the very foundation of our test and development environment for the rover in this book.

1. Install Oracle's VirtualBox from the following link:

   `www.virtualbox.org/wiki/Downloads`

   Go to the list of **host** operating systems and choose Microsoft Windows and install the VirtualBox software (orange box in Figure 3-1). For other operating systems, see `www.virtualbox.org/manual/ch02.html`.

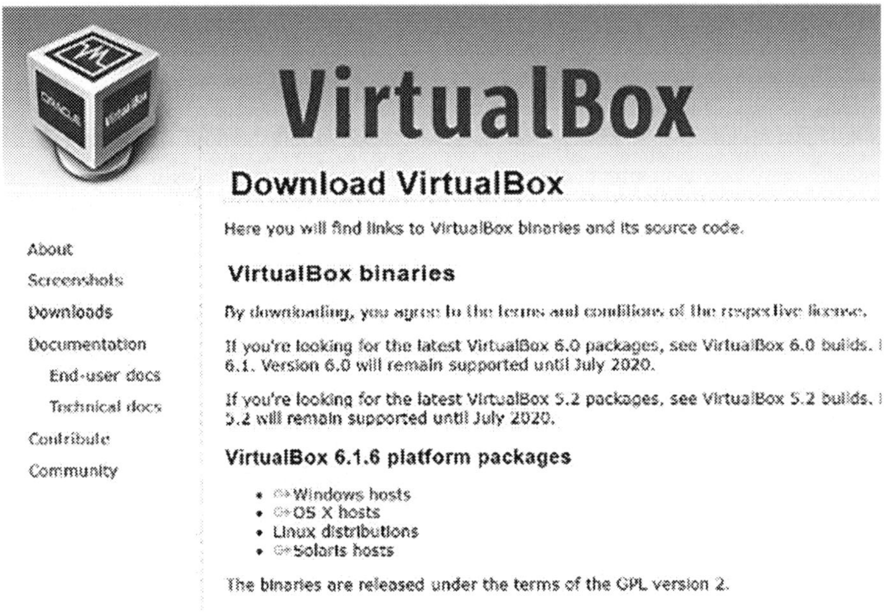

Download VirtualBox

Here you will find links to VirtualBox binaries and its source code.

About

Screenshots

Downloads

Documentation

   End-user docs

   Technical docs

Contribute

Community

VirtualBox binaries

By downloading, you agree to the terms and conditions of the respective license.

If you're looking for the latest VirtualBox 6.0 packages, see VirtualBox 6.0 builds. |
6.1. Version 6.0 will remain supported until July 2020.

If you're looking for the latest VirtualBox 5.2 packages, see VirtualBox 5.2 builds. |
5.2 will remain supported until July 2020.

VirtualBox 6.1.6 platform packages

- ⇒ Windows hosts
- ⇒ OS X hosts
- Linux distributions
- ⇒ Solaris hosts

The binaries are released under the terms of the GPL version 2.

Figure 3-1. Downloading Oracle's VirtualBox

2. After downloading the VirtualBox executable,
    find the file and double-click on it to execute. You
    should now see an image something like Figure 3-2.
    Click *Next.*

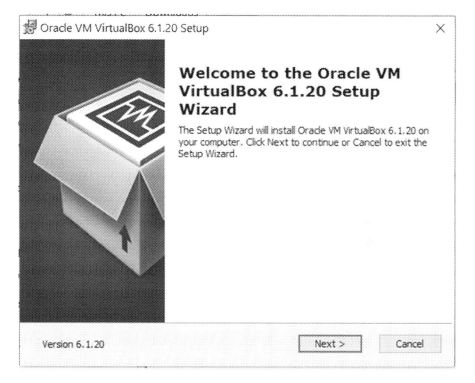

Figure 3-2. *Installing Oracle's VirtualBox*

3. Accept the default settings on the next couple of
    pop-up windows until you get the warning about
    resetting the network (Figure 3-3). Click *Yes*, and the
    VirtualBox program will install.

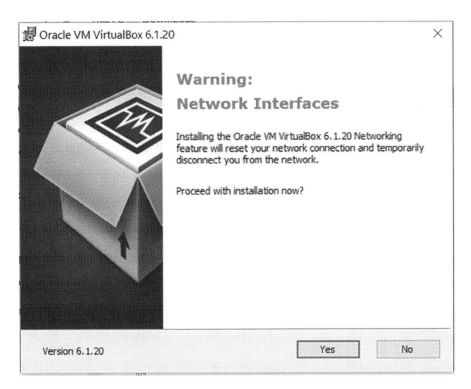

Figure 3-3. *Installing Oracle's VirtualBox (continued)*

Creating a New VirtualBox Virtual Machine

Now that we have installed Oracle's VirtualBox, run the program by double-clicking on its icon.

1. After it starts, you will see an area on the window that looks like Figure 3-4. To create a new machine, i.e., one that has not been installed before, click on *New*. If you do not see this ribbon bar, then you can access it from the ***Tools ➤ New*** toolbar.

Preferences | Import Export | New   Add

Welcome to VirtualBox!

Figure 3-4. *Creating a virtual machine within Oracle's VirtualBox*

2. Fill out the installation form like in Figure 3-5. Click
    Next. This does not install the operating system; it is
    a placeholder for later.

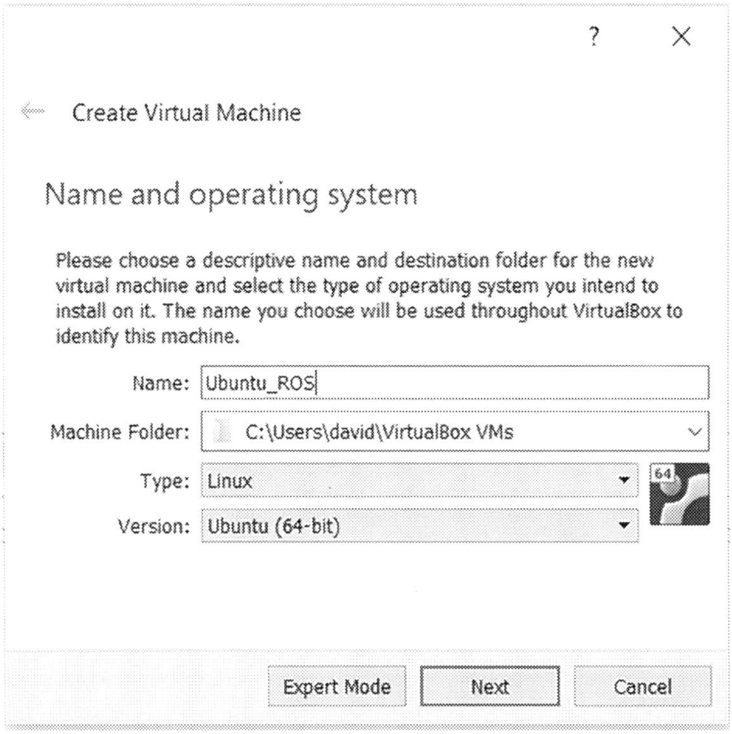

Figure 3-5. *Creating a virtual machine within Oracle's VirtualBox*

3. Next, we allocate memory for our development
    system. The default is 1024 MB, but we will allocate
    at least 2048 MB (Figure 3-6). You may allocate
    more, but 2048 MB is enough.

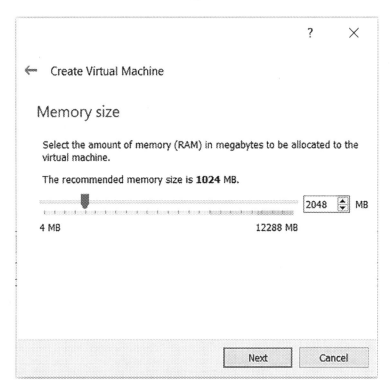

Figure 3-6. *Allocating memory for the virtual machine*

4. To create the virtual disk on the physical hard disk,
    accept the default settings shown in Figure 3-7.
    Since we do not have an existing drive, we have to
    create a new drive.

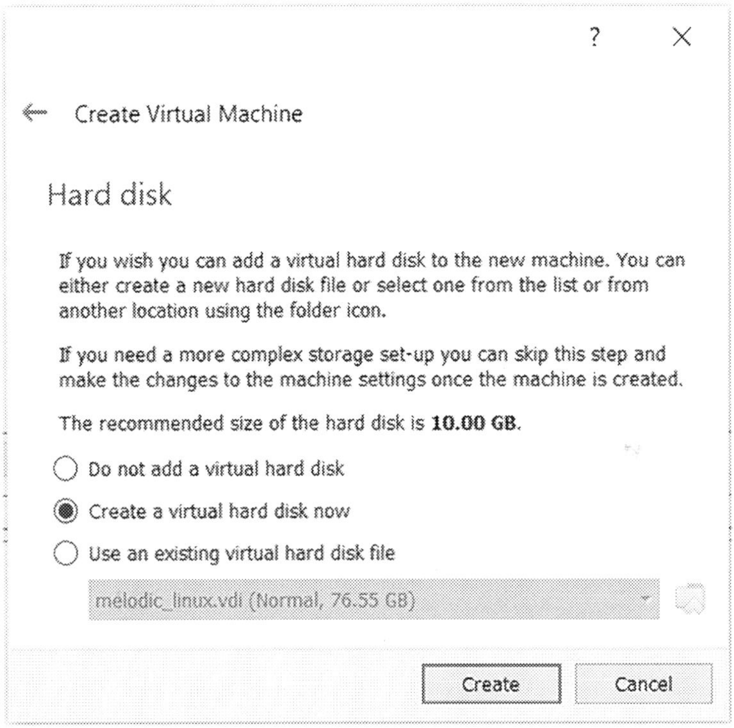

Figure 3-7. *Creating virtual hard disk*

5. Accept the default setting for the virtual hard drive
    type shown in Figure 3-8. We chose VDI because it is
    optimized for VirtualBox.

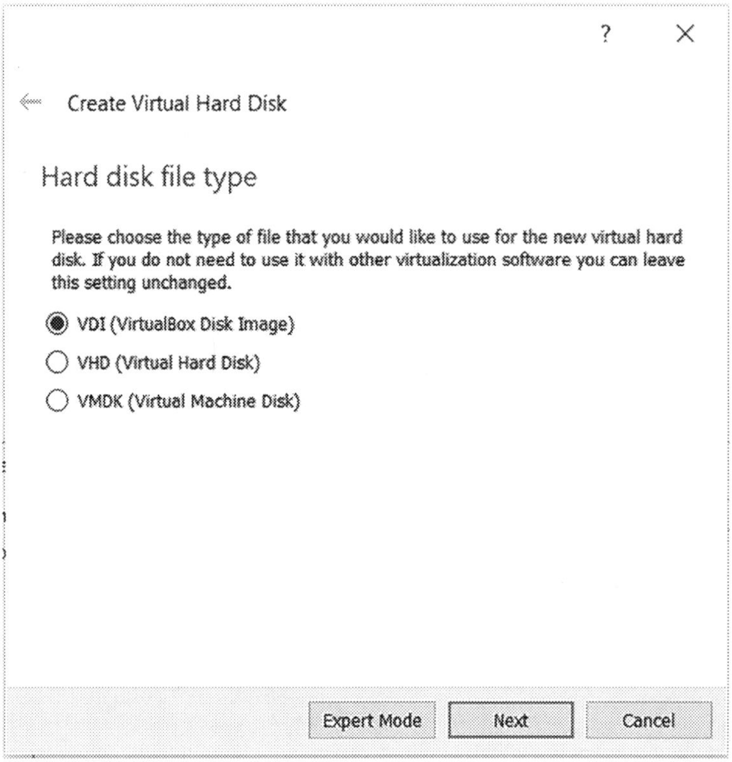

Figure 3-8. *Setting the virtual hard drive type*

6. Accept the default setting for the type of physical
    disk storage type shown in Figure 3-9. We
    chose dynamic, so the size can grow and shrink
    automatically.

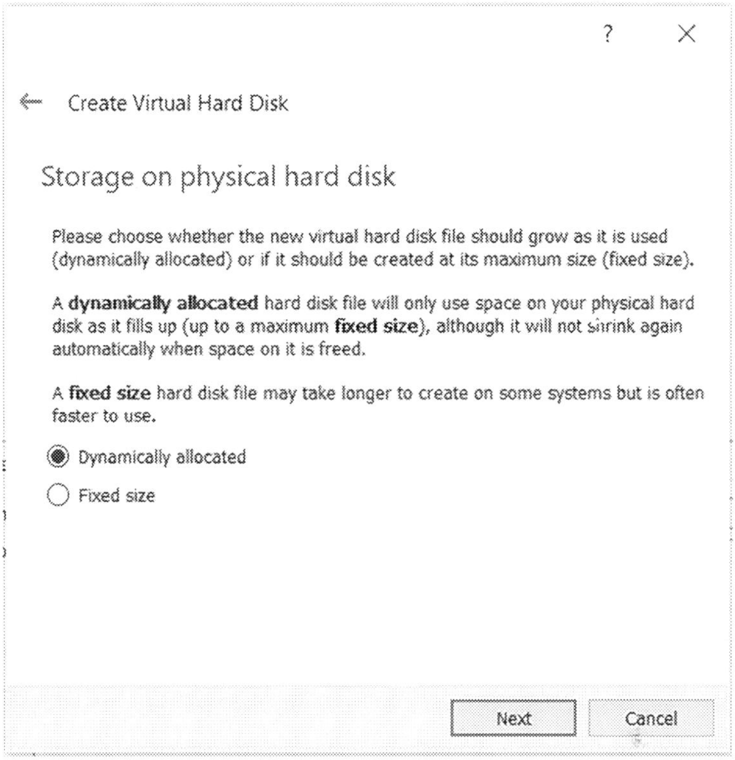

Figure 3-9. Setting the physical hard drive to allocate dynamically

7. Finally, set the location and maximum size your
   virtual disk can hold (Figure 3-10). The default
   location is derived from your original disk name,
   and the default size is okay for our little project.
   Click *Create.*

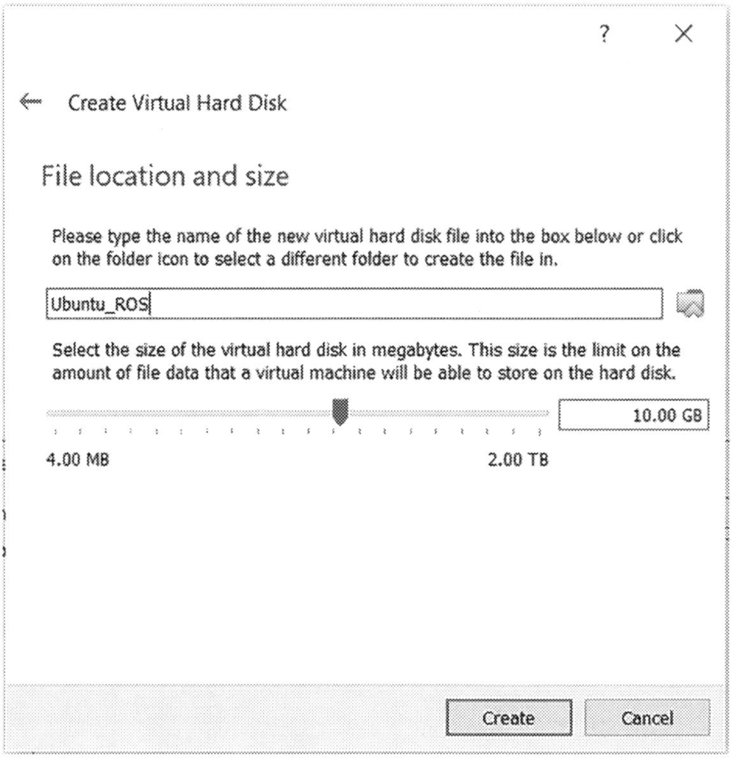

Figure 3-10. *Setting the location and size of the virtual disk*

8. You should now have a virtual machine installed, as shown by the orange box in Figure 3-11.

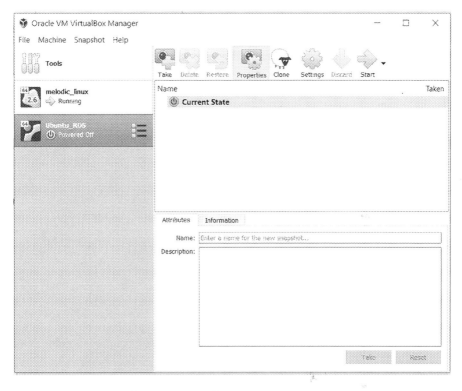

Figure 3-11. *Installation complete*

What we have just done is create an empty disk that we will next install in our Ubuntu Linux operating system.

Installing Linux Ubuntu 20.04.4 in VirtualBox

To download the required DVD/CD ISO image for Ubuntu Linux 20.04.4, a.k.a Focal Fossa, go to `http://old-releases.ubuntu.com/releases/20.04.4/`.

By default, the image is a 64-bit architecture, so you will see a 64-bit PC (AMD64) desktop image. If your physical RAM is less than 4 GB you should use the 32-bit architecture version. Even though it says "AMD" it is also the download for Intel chips.

1. After downloading the Ubuntu image, we need to load it into our virtual machine on VirtualBox. Click on Ubuntu_ROS to highlight that virtual disk. Click on Settings, and you will see a popup menu similar to that in Figure 3-12.

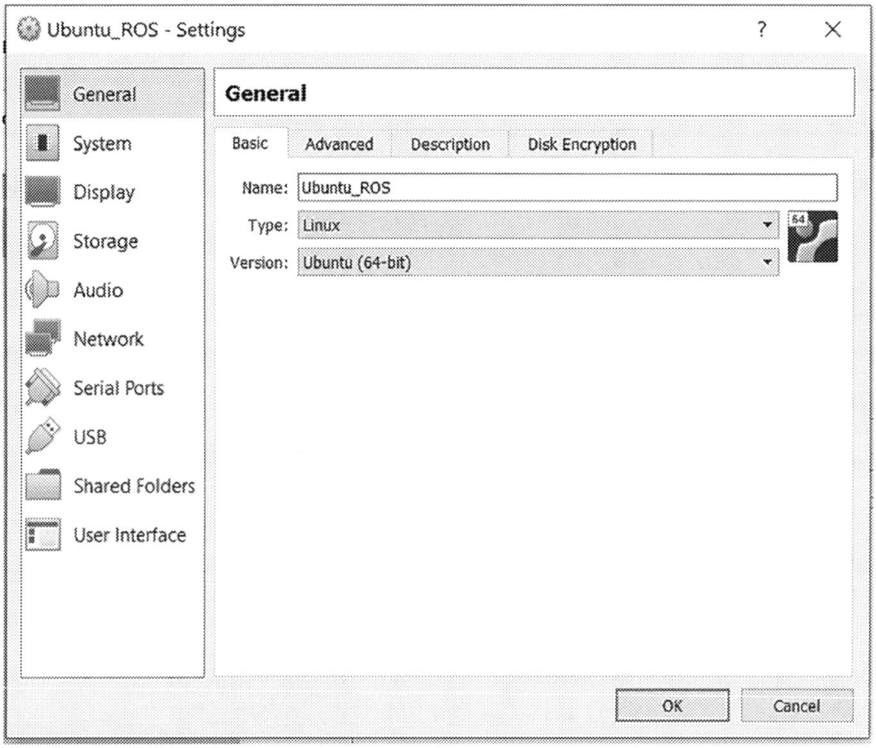

Figure 3-12. *ISO settings*

2. We need to install the specific version of the operating system that we set as a placeholder earlier ("Installing VirtualBox," Step 3). Our version is dictated by the Noetic ROS, which is specifically designed for Ubuntu Linux 20.04.4. Click on **Storage**. In Figure 3-13, notice the empty disk

icon (orange box) and the disk on the far right (red box). This is where we will attach our downloaded ISO image. Click the red box icon to connect the downloaded ISO file. Choose the first option, "Choose Virtual Optical Disk File..."

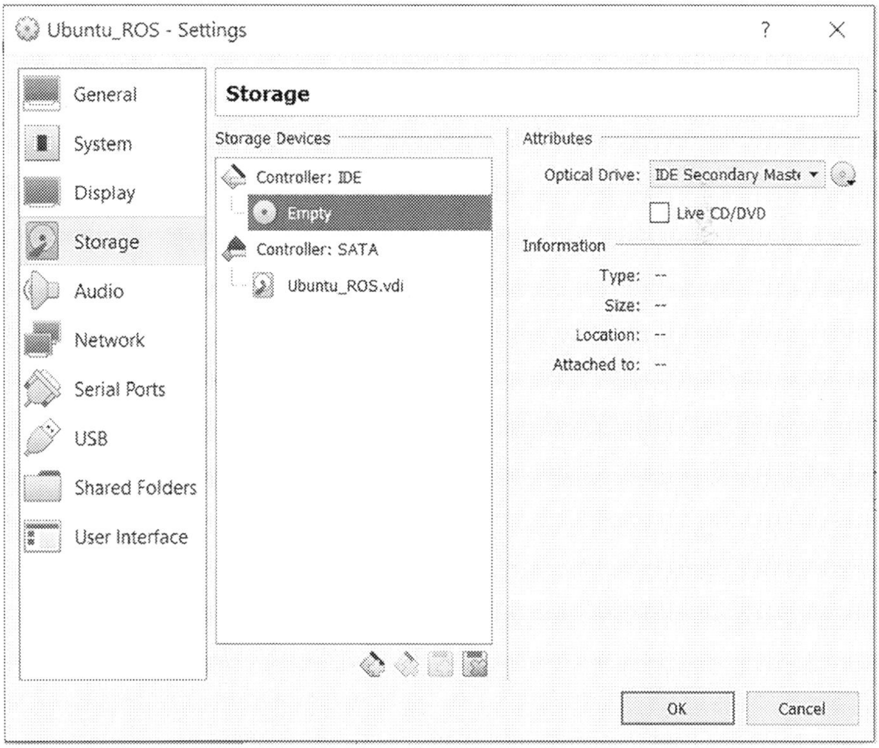

Figure 3-13. *Storage settings needed to attach ISO image*

3. Find and select your downloaded ISO file. Ours
   was saved in the "Download" folder. Select
   "Ubuntu-20.04.4-desktop-amd64.iso" and then click
   Open. This will attach the ISO file to your VirtualBox
   as the operating system. If you wanted to, you could
   now run your Linux system. However, we are going
   to tweak the settings of our operating system to
   make the development environment easier to use,
   as seen in Figure 3-14.

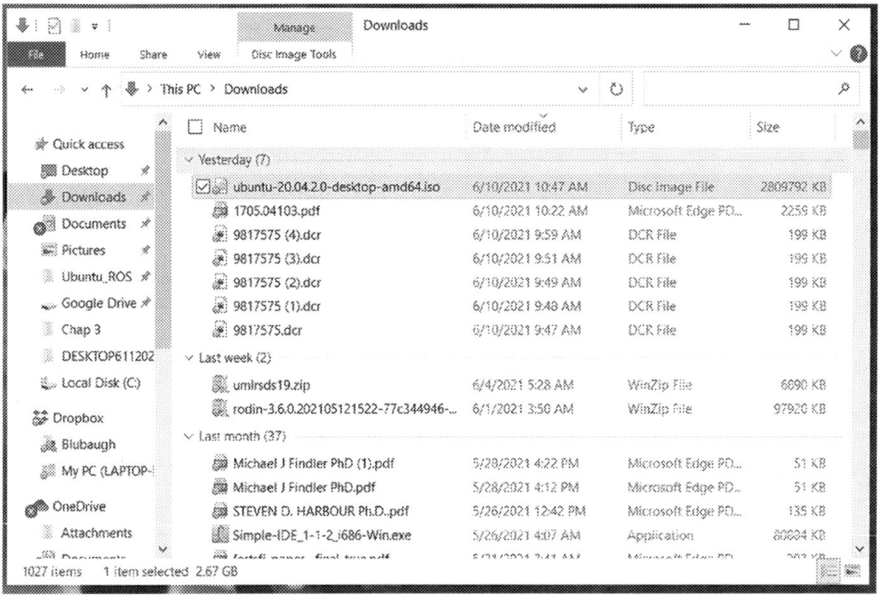

Figure 3-14.   *Select your ISO image and open it*

4. In Figure 3-15, the word "empty" has been replaced
   with your ISO image name. Your operating system is
   installed.

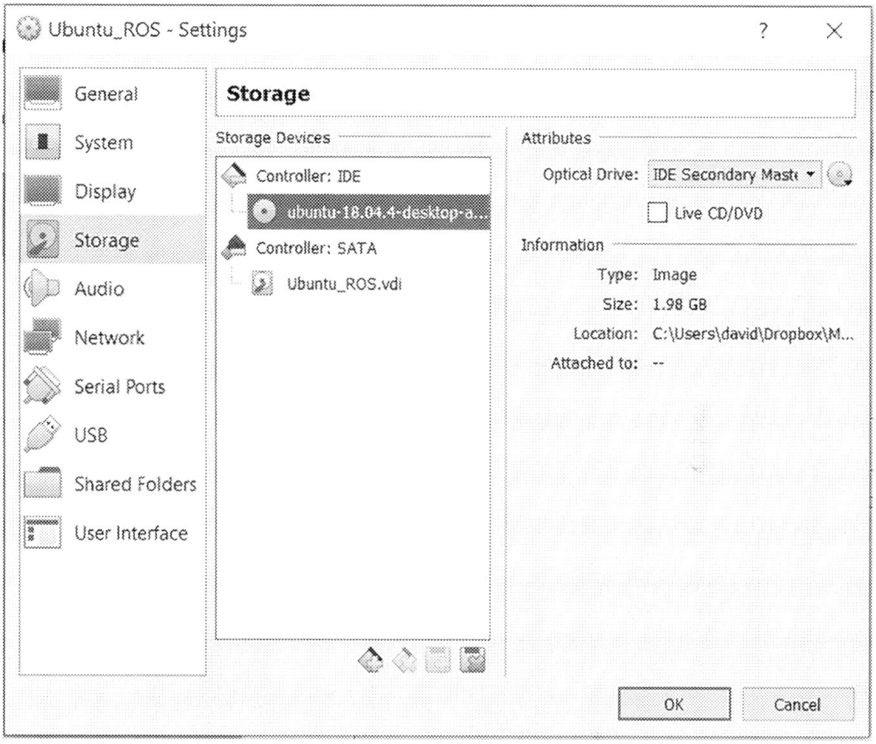

Figure 3-15. *ISO image attached*

5. Change your display settings to match Figure 3-16.
    We assume you have at least 4 GB video and
    SVGA. Setting the Scale Factor to 200% lets the
    window be a reasonable size.

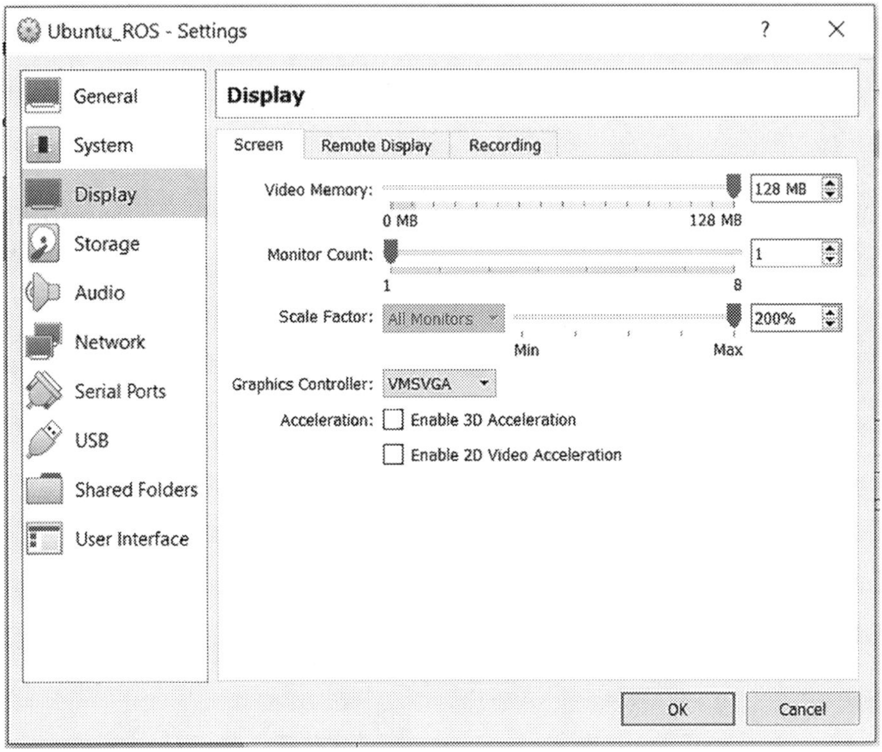

Figure 3-16. *Display settings*

6. Finally, we need to set a couple of system settings. Figure 3-17 shows the optimal RAM to be assigned (green bar), and Figure 3-18 shows the optimal CPUs that can be assigned (green bars). The green bars match your computer's hardware our system has 8 GB RAM and 4 CPUs, so that is what we assigned. If you choose any values in the red, then the system will be configured to a different set of hardware; therefore, it will be slower.

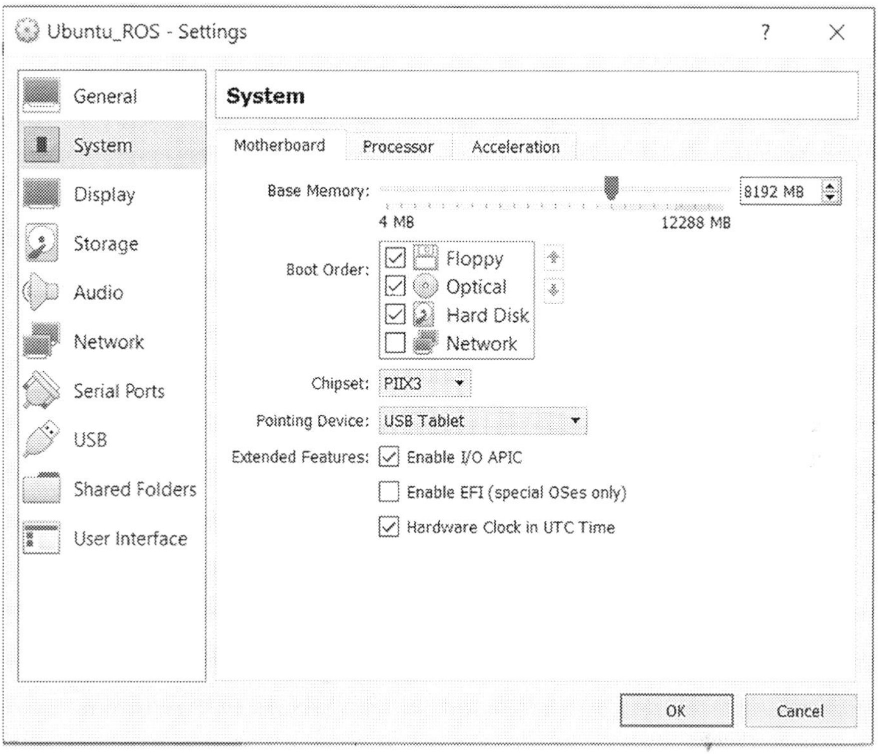

Figure 3-17. *System motherboard settings*

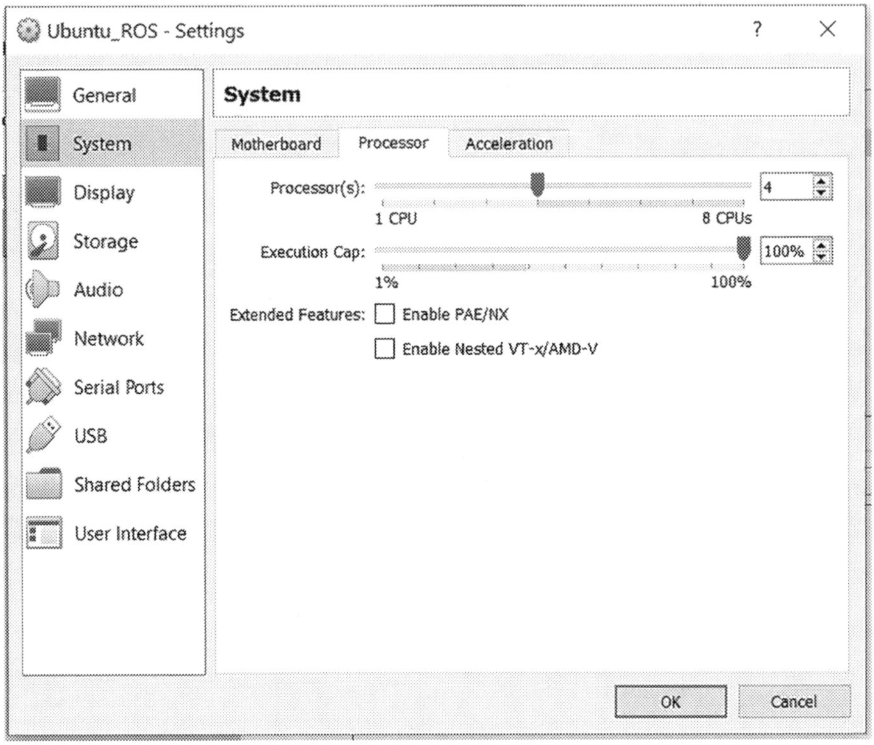

Figure 3-18. *System processor settings*

7.   Your system should now be ready to install a fully
     operational virtual Linux system. Click on *Start* (or
     double-click on your drive) to start installing your
     Linux system. See Figure 3-19.

Figure 3-19. *Starting the Ubuntu Linux OS as a guest
virtual machine*

8. Click on *Install Ubuntu* (Figure 3-20). We will be selecting the defaults most of the time during the installation. If necessary, change the settings for your installation.

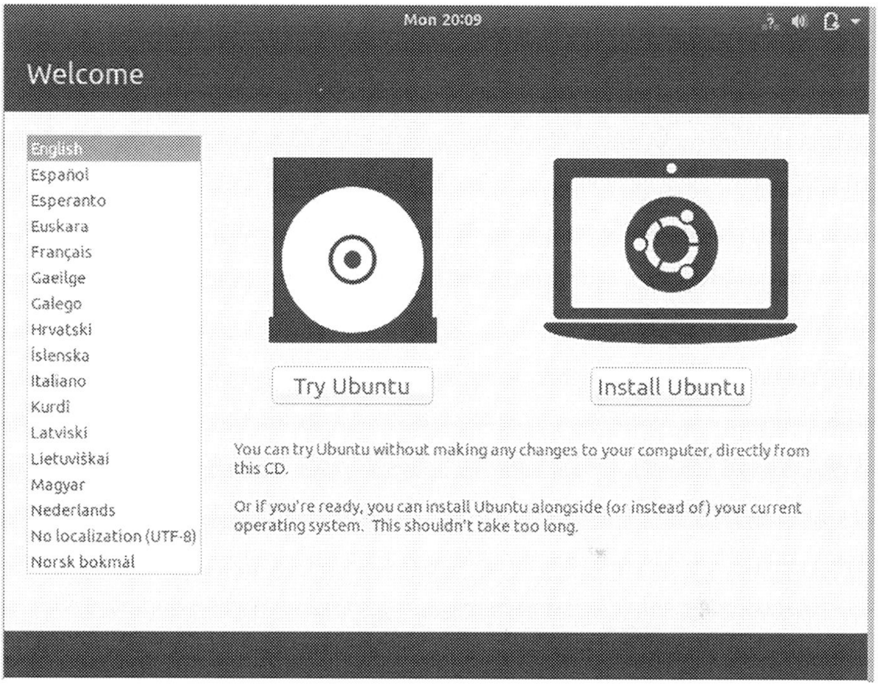

Figure 3-20. *Ubuntu Linux OS install setup*

9. Our next step is to set the keyboard preferences (Figure 3-21). The default settings were sufficient for us since we are currently located in the United States.

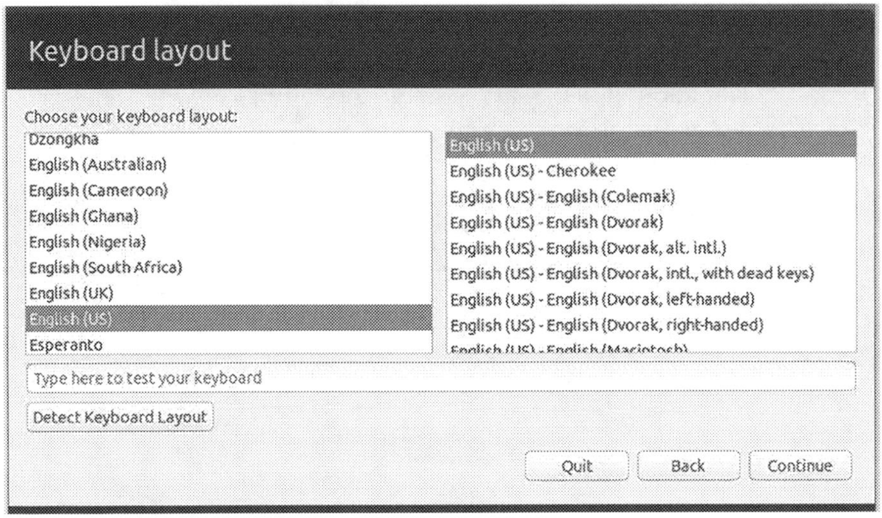

Figure 3-21. *Ubuntu keyboard preferences*

10. Once we have established the keyboard settings,
     we can then set the normal installation for Ubuntu
     (Figure 3-22).

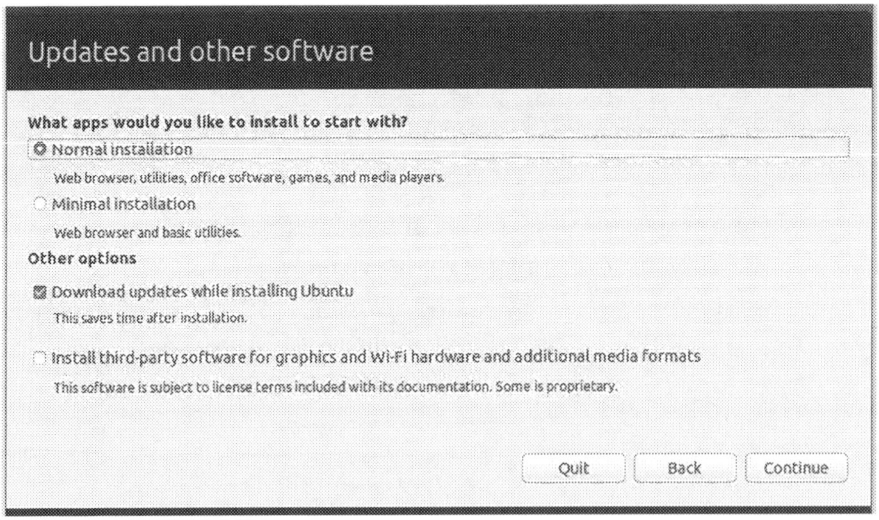

Figure 3-22. *Normal Ubuntu installation*

11. We assume there are no prior installments of
     Ubuntu before proceeding with the final installation
     (Figure 3-23). The default settings are once again
     sufficient for us.

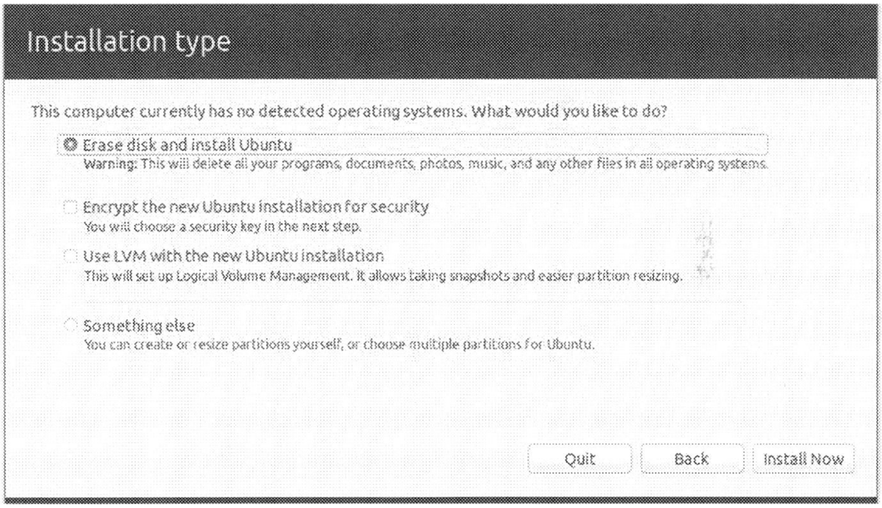

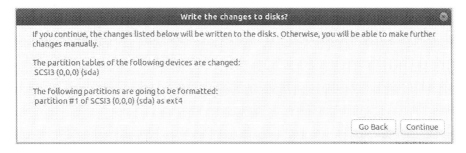

Figure 3-23. *Erase disk and install Ubuntu (continued)*

12. Once the installation process begins, we might also
     encounter a warning that declares that we are about
     to rewrite the targeted virtual disk (Figure 3-24). We
     then click *Continue*.

Figure 3-24. *Ubuntu installation warning*

13. Set the correct time zone for the Ubuntu computer
     system (Figure 3-25).

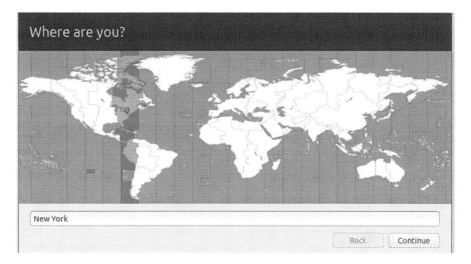

Figure 3-25. *Ubuntu timezone setting*

14. Finally, we input the name, username, and
     password for the Ubuntu system (Figure 3-26).
     Since we are only using Ubuntu as a development
     platform, we should use a strong password for
     security purposes. To maintain the highest levels
     of password security for the Ubuntu Development
     environment, we could use numbers 0–9, letters a–z
     or A–Z, and special characters such as #, ^, and *.

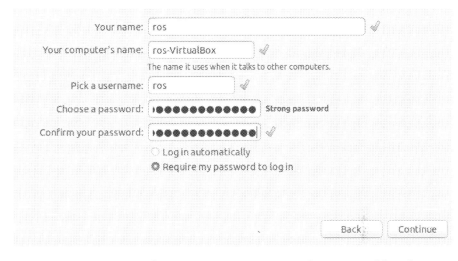

Figure 3-26. *Setting the name, username, and password for the Ubuntu installation*

Congratulations! You have successfully installed Ubuntu Linux, as indicated by the green rectangle in Figure 3-27. Next, we will need to install the tools needed for software development. Ubuntu gives us Python as our programming language and Gedit text editor. We need to install ROS, which will give us the library to control the rover. The full ROS installation will give us simulator and visualization tools, Gazebo, and Rviz. We will add a couple of other libraries that will be used later in the book, namely TensorFlow (deep learning) and OpenCV (computer vision).

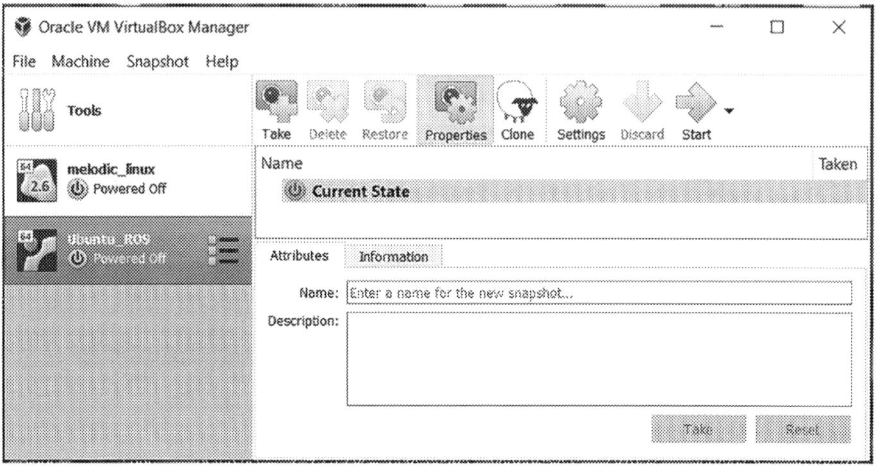

Figure 3-27. *Ubuntu finally installed*

Updating Ubuntu Linux 20.04.4

Now that we have installed Ubuntu, to start the operating system, click on the *Start* button (Figure 3-27). You may be prompted for additional installation information; just reply "yes" to the default choices. It will also ask you to log in. After the operating system launches, you should have a window that looks like Figure 3-28. (The VirtualBox menu and bottom of the screen commands will be cropped off of future images.) We are now ready to install ROS through terminal commands.

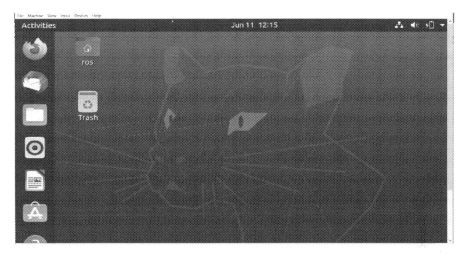

Figure 3-28. *Ubuntu initial screen*

In Figure 3-28, the desktop includes system application icons on the left, a trashcan to delete items, a header bar that has the day and time, and four icons in the upper right-hand corner (network, sound, battery, and down arrow). We will discuss these as necessary.

1. To start a terminal window, right-click on the desktop background, being sure to avoid icons (Figure 3-29). Click on the **Open in Terminal** menu choice.

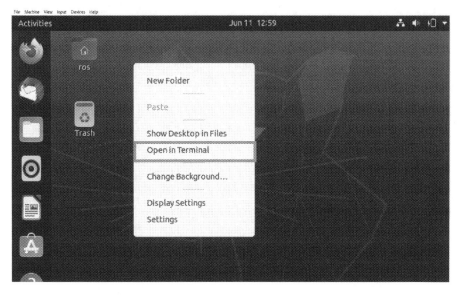

Figure 3-29. *Ubuntu open terminal selection. Access menu by right-clicking*

2. On a freshly installed Linux system, your terminal will be full screen (Figure 3-30). Later on, we will run multiple terminals at the same time, so let's make the terminal "windowed" by clicking on the □ in the upper-right corner of the terminal, as indicated by the orange arrow.

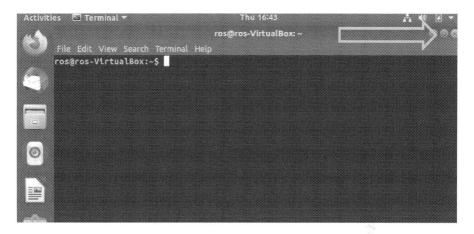

Figure 3-30. Ubuntu terminal session

3. Now we have a terminal window on our desktop
    (Figure 3-31). The installation of ROS will be done
    through this terminal window using commands. We
    are not going to fully explain the commands, since it
    is beyond the scope of this book, but simple internet
    research should suffice if you want to investigate
    further.

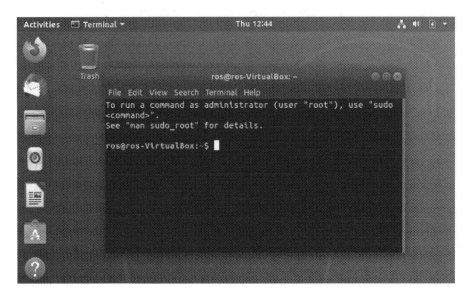

Figure 3-31. *Ubuntu initial terminal screen*

4. Even though we just installed Ubuntu Linux
    20.04.4, there may be software updates to drivers,
    applications, etc. We need to update the operating
    system. We will be using sudo apt-get commands.
    Sudo means "superuser do" and it temporarily
    elevates you with system administrator privileges;
    therefore, you will have to enter your password.
    Execute the following command in the terminal to
    update the necessary features:

    ```
    $ sudo apt-get update
    ```

    We can also see the first command in Figure 3-32
    highlighted in the orange rectangle. After entering your
    password, you will see a few lines of updates. This is a
    download of lists of the latest version numbers of operating
    system files, not the upgrades! Enter the following:

```
$ sudo apt-get upgrade
```

to upgrade your operating system files to the latest versions.

After it stops, you will now have the newest version of Ubuntu Linux 20.04.4 installed (Figure 3-33). As an aside, for this chapter, from now on, we will show commands in the terminal in an orange rectangle, not in text prose.

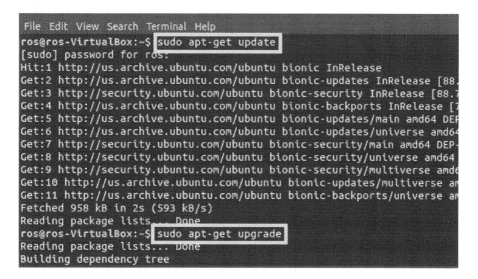

Figure 3-32. *Ubuntu terminal update command*

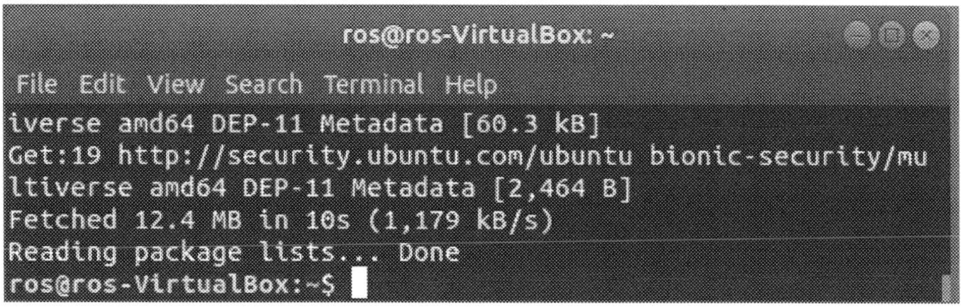

Figure 3-33. *Ubuntu terminal update completed*

We have now updated our Ubuntu Linux system to the latest version. Next, we need to set up Ubuntu to allow external libraries (ROS, etc.) These libraries come from external repositories.

Configuring Ubuntu Repositories

To download repositories from external third-party software sites, such as ROS.org, we need to tell the system about these sites. To do this, click on the System Settings icon on the left side of your desktop (Figure 3-34).

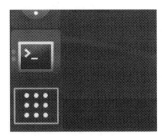

Figure 3-34. *Ubuntu system settings icon (orange box)*

Up pop several icons. Click the Software Updates icon on the desktop (Figure 3-35).

Figure 3-35. *Ubuntu software update icon*

Next, we select the Settings option as in Figure 3-36. The Settings option defines the software to be installed or upgraded, including third-party software, such as ROS.

Figure 3-36. *Ubuntu software settings option*

On the new popup window, select the **Ubuntu Software** tab and set the checkboxes shown in Figure 3-37. Once those selections have been made, then please click *Close*. These selections approve the download of third-party software.

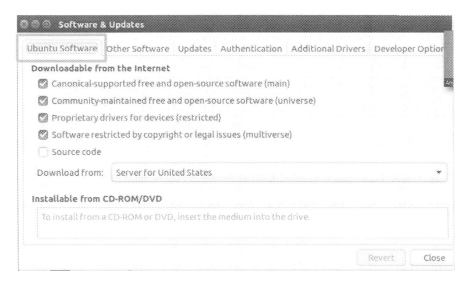

Figure 3-37. *Display of selected repository checkboxes*

Installing Anaconda

As a reminder, the ORDER of installation is important: VirtualBox, Ubuntu, Anaconda, ROS. Any other order may cause problems.

Developing a complex software project is made more manageable if we use an integrated development environment (IDE). An IDE combines a simple text editor (Gedit) and a compiler (Anaconda). Doing the programming in an IDE makes the programming process easier because the built-in tools and the IDE are smart enough to catch simple errors. There are many great IDEs available for Linux; we chose Gedit for its elegant simplicity. The Anaconda compiler can run Python 2.7, which is required by ROS.

First, we need to install the Anaconda Python compiler from `https://anaconda.com/products/individual`. Scroll down to find the link for the 64-bit Linux download (Figure 3-38).

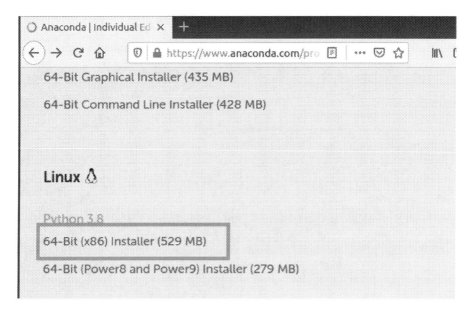

Figure 3-38. *Anaconda install*

Now that we have downloaded the installation file, we need to locate it. Figure 3-39 shows the four steps: 1) click on the file cabinet; 2) click on downloads; 3) you will see the Anaconda shell script, and 4) open a terminal at this disk location using a right-click near the file name.

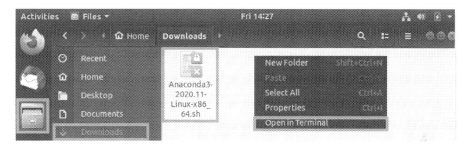

Figure 3-39. *Anaconda locating download*

After opening the terminal, enter the two commands shown in Figure 3-40. The first command makes the shell script executable. The second command executes the shell script so that it is installed on Linux.

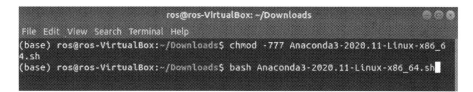

Figure 3-40. *Anaconda bash install*

Now that we have installed our compiler, we need to set up the operating system to accept ROS. We do this by setting up the ROS source list and key.

ROS Source List

Now that we have successfully configured our Ubuntu repositories to allow for "restricted," "universe," and "multiverse" repositories, allowing the installation and updating of the necessary supporting software for ROS, we can now set up the required software list from the ROS.org Foundation website. This can be accomplished with the following terminal command:

```
$ sudo sh -c 'echo "deb http://packages.ros.org/ros/ubuntu
$(lsb_release -sc) main" > /etc/apt/sources.list.d/ros-
latest.list'
```

The ROS source list has now been configured. Now we will set the ROS environment key.

ROS Environment Variable Key

The ROS environment variable key verifies and validates the origin of the ROS source-code repository. It also guarantees the software has not been modified without the consent of the owner. The key makes the ROS repositories "trusted" software sites for downloads, installations, and updates. Type in the following shell command at the terminal prompt (Figure 3-41):

```
$ sudo -E apt-key adv --keyserver 'hkp://keyserver.ubuntu.
com:80' --recv-key C1CF6E31E6BADE8868B172B4F42ED6FBAB17C654
```

Figure 3-41. *Getting Noetic ROS install key*

Note   The ROS.org Community recently discovered a security flaw with the original key for Noetic ROS. That is why we are using this newer key. The reader needs to be vigilant of all security upgrades, maintenance, and/or additional enhancements within the ROS ecosystem. The reader needs to occasionally review and check for such security updates within the ROS.org discussion forums.

Installing the Robotic Operating System (ROS)

The two commands needed to install ROS are shown in Figure 3-42. The first command is just good practice and verifies all the system software is up to date, while the second command installs the full ROS library. The downloading and installation of over 1,000 files can take a long time. (My computer took more than 30 minutes.)

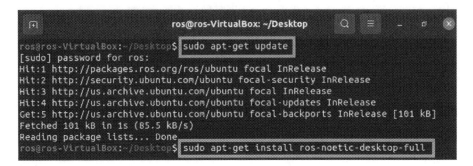

Figure 3-42. *Ubuntu ROS install*

The full-featured version of Noetic ROS should now be installed. Let us take this system out for a spin.

Note   If you receive an error that none of the Noetic packages could be located, you might be required to upgrade the Ubuntu software. Please execute the following terminal commands:

$ sudo apt-get update

$ sudo apt-get upgrade

Installing ROSINSTALL

You have successfully installed ROS to your Linux system. However, there exist some additional libraries that will make the development of your project simpler. These are accessed through rosinstall:

$ sudo apt-get install python3-rosinstall

$ sudo apt-get install python3-rosinstall-generator

$ sudo apt-get install python3-wstool build-essential

Be very careful about the spaces and dashes in the previous command. If you get an error, verify every dash and space.

Starting ROS for the Very First Time

The next three commands are used to wake up ROS and check for any ROS-dependent library updates. These should be run the first time you start up ROS. The ROS we downloaded and installed in Figure 3-42 may not have the latest subcomponents. The second command checks for any updates available for subcomponents.

```
$ sudo apt-get install python3-rosdep
$ sudo rosdep init
$ rosdep update
```

Adding the ROS Path

Every time you want to use ROS, you need to tell the operating system where to look for the ROS commands, i.e., add the ROS path to the global path. Since we have to do this every time we want to use ROS, we will create a bash script. In a text editor (Gedit), create a file called STARTrosconfig.sh with the included lines (Figure 3-43).

Figure 3-43. Starting ROS configuration in Gedit

Next, save the file to the /home/ros directory. This will be reflected in the Gedit user interface by the small orange box, as in Figure 3-44. Open a new terminal. Using Figure 3-38 as our guide, we are going to make STARTrosconfig.sh executable. Orange boxes are commands you will type in, and green boxes are outputs you should observe. After you type in your first two commands, observe that STARTrosconfig.sh is white, which means it is not executable. The next command (ls -la STARTrosconfig.sh) is optional and looks at the details for the file. You should note that the file can be read (r) and written (w), but not executed (-). To change this, you must issue the change mode command (chmod +rwx STARTrosconfig.sh). To verify the change occurred, ls -la STARTrosconfig.sh will show rwx and a green name. Success!

Figure 3-44. *Making STARTrosconfig an executable*

Every time you want to work with ROS, open a terminal and type source STARTrosconfig.sh. This adds the ROS path to the system (Figure 3-45). To verify the ROS variables were installed correctly, type env | grep ROS_*. This command looks at the environment variables (env) and selects only those variables starting with "ROS_".

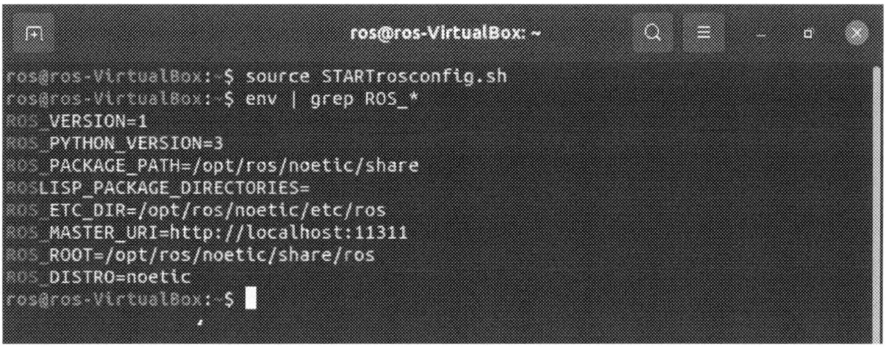

Figure 3-45. *Verify ROS paths*

Creating a ROS Catkin Workspace

Once the ROS environment variables have been established, we create a workspace directory to develop the ROS applications. ROS requires a "catkin" workspace consisting of three subfolders (src, build, and devel). Each of these folders needs script files to help compile the project. This workspace should be located in the user home directory. Therefore, use the following shell commands to create the workspace:

```
$ cd ~
$ mkdir -p ~/catkin_ws/src
$ cd ~/catkin_ws/src
$ catkin_init_workspace
$ cd ..
$ catkin_make
```

Note   If you receive the following error—"Ackermann msgs were
not found by Cmake"—or that any other type of error message was
not found, then you need to manually install the correct package,
such as the Ackermann package, as follows:

```
$ sudo apt install ros-noetic-ackermann-msgs
```

The script command `catkin_init_workspace` sets the `src` folder as the
project source file repository. The script command `catkin_make` creates
the `devel` and `build` folders and adds the necessary script files. We will use
the `catkin_ws` workspace as the central directory to develop and test all of
the software components. Now we need to make certain that the `catkin_
ws` workspace is sourced for ROS:

```
$ source ~/catkin_ws/devel/setup.bash
$ echo "source ~/catkin_ws/devel/setup.bash" >> ~/.bashrc
```

Now you should have the `catkin_ws` workspace successfully sourced
as the main workspace that operates on top of the ROS environment. As
such, if you execute the following shell text command:

```
$ echo $ROS_PACKAGE_PATH
```

It should then display the following output to the terminal
prompt window:

```
/home/<username>/catkin_ws/src:/opt/ros/noetic/share
```

Note   <username> will be the Ubuntu Linux OS username you
defined earlier.

The catkin_ws workspace has now been sourced as the main workspace directory for developing applications within the ROS environment. If we did not obtain the ROS package path directory output as expected, please consult debugging issues within the ROS.org Community website for further information.

Final Checks for Noetic ROS

The following shell text command will determine if the correct version of ROS has indeed been installed and is operational:

```
$ rosversion -d
```

If the output is Noetic, then you have completely succeeded in obtaining a fully operational ROS environment. Congratulations.

Note   All shell commands used in this book are compatible with both the current and future Noetic versions of ROS. The only difference might be the number of parameters and substituting the word "Noetic" for the correct rosversion where necessary.

Noetic ROS Architecture

With the successful installation of Noetic ROS, we will introduce its architecture. The ROS architecture was designed to operate with robots of all types (rovers, drones, planes, boats, submarines, etc.). As such, the ROS framework was designed to handle multiple components or nodes in a robot. A node is a sensor, motor, controller, etc. In other words, it is a piece of hardware that performs a function for the robot. ROS refers to a "program node" as the software necessary to control the hardware. The primary node is known as the central master node.

The ROS architecture handles the communication between nodes. For example, if the central master node receives data from a distance sensor node indicating an object is in our path, then the master node might send move commands to a driver node to change the robot's directions (Figure 3-46). The ROS architecture helps us to manage the increasing complexity of a robot system as more features, sensors, and actuators are added.

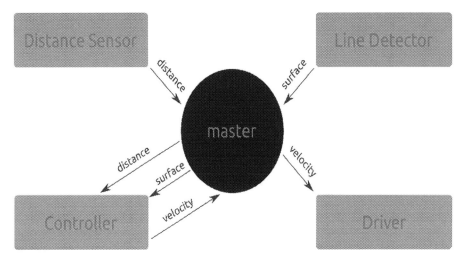

Figure 3-46. *Preliminary ROS architecture AI rover layout*

Figure 3-46 shows five nodes: the master node and four system nodes. When a system node starts, it sends information to the master comprising the type of data that the node can send and receive. The nodes that send data to the master node are called *publisher nodes*. For example, the Distance Sensor and Line Detector nodes are publisher nodes, since they send data (distance to target and surface information, respectfully) to the master node. The nodes that receive data from the master node are called *subscriber nodes*. So the Driver node is a subscriber node since it receives velocity data from the master node. The Controller node is both a publisher and a subscriber node.

The Controller node will be hosting the deep learning and cognitive deep learning routines developed in later chapters. These deep learning routines will send and receive messages to and from the ROS master node. The master node will reroute that message to their intended nodes, such as a message to the Driver node to control the speed.

Simple "Hello World" ROS Test

Now that we have described the basic ROS architecture, let's determine if our installation of ROS was successful by executing simple ROS tutorial scripts. This script will simply have two ROS nodes: a talker (publisher) and a listener (subscriber). If we look at Figure 3-47, we see three nodes. First, when the listener is created, it tells the master node that it wants to listen to the talker (subscribes). Next, when the talker is created it registers with the master node. Finally, the talker generates (publishes) a message; this means it sends the message to the master. This causes the master node to send the message to all nodes listening to the talker, which means the message is sent to the listener.

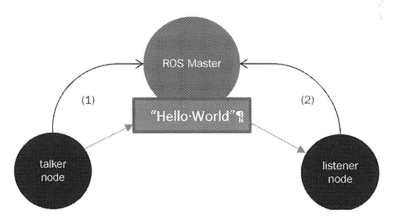

Figure 3-47. *Simple talker and listener node ROS test example*

We are eventually going to open three terminals, each one a node. (If you want to, install **Terminator** from the Ubuntu Software store to open multiple tiled terminals. It makes running the ROS scripts easier.) Open the first terminal, which we will make the ROS master node:

```
$ roscore
```

Go back to the desktop and open a second terminal. We will make it the listener:

```
$ rosrun roscpp_tutorials listener
```

Go back to the desktop and open a third terminal and make it the talker:

```
$ rosrun roscpp_tutorials talker
```

You should then see on the talker terminal:

```
hello world 0
hello world 1
hello world 2
...
```

And you should then see on the listener terminal:

```
I heard: [hello world 0]
I heard: [hello world 1]
I heard: [hello world 2]
...
```

If you have made it this far, congratulations, you have successfully installed VirtualBox, Ubuntu Linux, and Noetic ROS. (Do not shut down the nodes, as we will use them in the next section.) Next, we visit two components that come with ROS: RQT Graph and Gazebo Robot Simulator. (We will talk about Rviz and TensorFlow when we use them.)

ROS RQT Graph

The RQT Graph is an important visualization and debugging tool. This graphing tool will be used to debug running nodes and inspect the communication between them. We will use RQT Graph to inspect our two nodes that are running from the previous section. In a new terminal, enter the following command, and you will get something like Figure 3-48:

```
$ rosrun rqt_graph rqt_graph
```

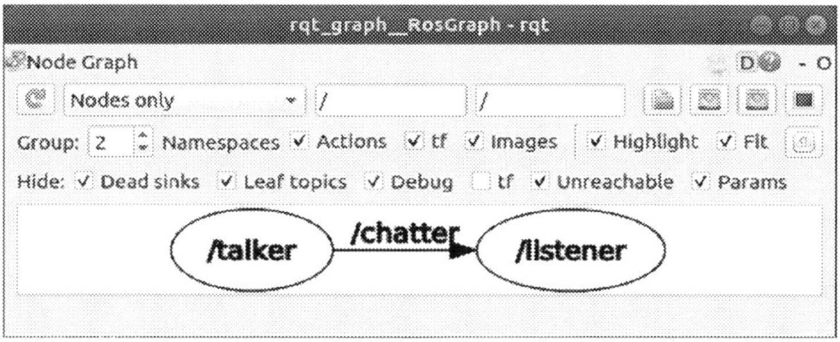

Figure 3-48.   *Talker node sending information to listener node*

The ROS command `rosrun rqt_graph rqt_graph` graphically displays all active publisher and subscriber nodes in an RQT Graph window. The interpretation of Figure 3-43 is that `/talker` is sending a message to `/listener`. Notice that the ROS master node is left out. The default message pipeline name is `/chatter`.

ROS Gazebo

Gazebo is a graphical simulation tool used to visualize any virtual world you create. The virtual world can contain objects, robots, buildings, obstacles, etc. You define them in terms Gazebo understands, and it will be simulated in a Gazebo window.

Gazebo is a standalone program that needs to be tied to the ROS system and world definition. This means we will "launch" the program from "within" ROS using the roslaunch command. Briefly, the ROS way of doing things: rosrun and roslaunch. The command rosrun launches a (Python) script to run an object's script by itself (or a limited number of other objects). In contrast, roslaunch loads and executes all objects (and associated scripts) in a world environment, where each object can interact with the other objects.

The roslaunch tool is the standard ROS method for starting Gazebo's simulated world and robots. To verify that Gazebo is correctly installed, we will launch the supplied willowgarage_world.launch shown in Figure 3-49 with the following command:

```
$ roslaunch gazebo-ros willowgarage_world.launch
```

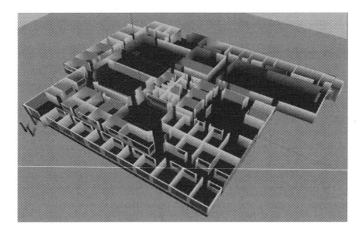

Figure 3-49. *Willow Garage generated world*

Now that we have verified Gazebo is installed correctly, we will be replacing the willowgarage_world.launch file with our world: a simplified simulated world of the unexplored Egyptian Catacombs and our intrepid AI rover. The AI rover's scripts with adaptive intelligence and decision making will circumvent ancient dangers that might still await us.

Summary

To start our project, we installed VirtualBox and a virtual Ubuntu Linux 20.04.4 LTS operating system. We then installed Anaconda, our Python programming interpreter, on Ubuntu. Lastly, we successfully installed the Noetic ROS on Ubuntu. This completed setting up our development environment.

To verify the environment was successfully installed, we created and ran two ROS nodes that communicated through Python scripts. We used RQT Graph to visualize the running nodes. Finally, we verified the Gazebo simulator could be launched from ROS. (I don't know about you, but I am tired.) Now that we have set up our development environment, we can start designing and developing our AI rover explorer and our simplified virtual Egyptian catacombs world.

EXTRA CREDIT

1. What other worlds can be explored within the `roslaunch` command for the Gazebo simulator? If need be, please consult internet resources.

2. What other tests could be used to determine a successful installation of ROS?

3. What other messages can be exchanged between nodes?

CHAPTER 4

Building a Simple Virtual Rover

After installing the development operating system (Ubuntu), the target operating system (ROS), and their associated tools in the last chapter, we are going to "play" with the tools to build a very simple rover with RViz and drive it in the Gazebo simulator. We will also build, test, and run the chassis of the rover one part at a time.

Objectives

The following are the objectives required for successful completion of this chapter:

- Understand the relationship between ROS, RViz, and Gazebo

- Expand your understanding of ROS commands

- Explore RViz to create a simple rover

- Use Gazebo to move the rover in a simple virtual environment

© David Allen Blubaugh, Steven D. Harbour, Benjamin Sears, Michael J. Findler 2022
D. A. Blubaugh et al., *Intelligent Autonomous Drones with Cognitive Deep Learning*,
https://doi.org/10.1007/978-1-4842-6803-2_4

ROS, RViz, and Gazebo

As a reminder, ROS stands for Robot Operating System. Our rover will use ROS as its operating system. An operating system is software that connects the different components of the system. The components can be hardware (motors, sensors, etc.), software (neural network libraries, etc.), or "squishy-ware" (humans). [Although the last term is added for humor, the operating system is key to interacting with the user.] RViz and Gazebo are software tools used in the development of an ROS robot (Figure 4-1). RViz is used to build models of our virtual rover. Another way to think about the two programs is that RViz explores individual object(s) in a controlled space (lab), while Gazebo puts the objects in a chaotic "real-world" environment with little control of interactions.

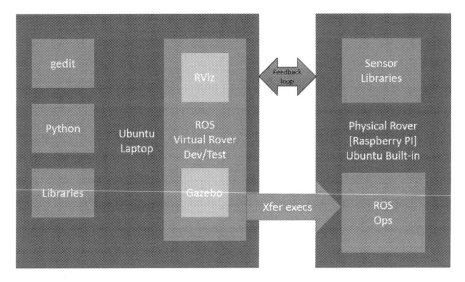

Figure 4-1. *Development system relationships*

Figure 4-1 graphically describes our project's major components and their interrelationships. The blue boxes are our physical computing systems (laptop and rover), which have Ubuntu operating systems. The orange boxes represent software components and libraries installed on each system. The yellow boxes are internal ROS tools for developing and testing ROS models. Once a virtual ROS model has been thoroughly vetted, the executable script is transferred to the physical rover (green arrow). Assuming everything is working, our rover will be able to move about in the real world and transmit data back to the laptop (gray arrow). The gray arrow represents the "human-in-the-loop" decisions that might be used to control the rover, such as "Start," "Come Home," or "Pause." Gazebo will allow us to view the effects of physics on the chassis, simulate the power applied to each motor, and simulate the algorithms.

Essential ROS Commands

Table 4-1 lists the ROS commands we will use frequently. These ROS commands allow us to control, analyze, and debug nodes contained in a package. A node is a self-contained model of a sub-part of the system (package). A package contains the different models we are using to describe our rover. For instance, our rover will have a model composed of a chassis, wheels, sensors, etc., which are the sub-models. Each of these sub-models may have physics models applied, such as speed and acceleration. Furthermore, our rover will interact with walls, holes, and obstacle models. All of these models and sub-models make up our rover package. The nodes mentioned in Table 4-1 usually correspond to a "physical" sub-model, such as the wheels.

Table 4-1. *Essential ROS Commands*

Command	Format	Action
roscore	$roscore	Starts master node
rosrun	$rosrun [package] [executable]	Executes a program and creates nodes
rosnode	$rosnode info [node name]	Shows information about active nodes
rostopic	$rostopic <subcommand> <topicname> subcom: list, info, & type	Information about ROS topics
rosmsg	$rosmsg <subcom> [package]/ [message] subcom: list, info, & type	Information on message types
rosservice	$rosservice <subcom> [service] subcom: args, call, find, info, list, and type	Runtime information being displayed
rosparam	$rosparam <subcom> [parameter]	Get and set data used by nodes

Rather than go into the details of each command, we will explore them more deeply when we use them in the text.

Robot Visualization (RViz)

The final model of our "simplified virtual rover" is composed of four sub-components (Figure 4-2): a chassis, a castor, and two wheels. The different components are modeled as a box, sphere, and disks. I think one of the key takeaways here is that the model does not have to look like the physical rover. To quote British statistician George Box, "All models are wrong, but some are useful." We have a very useful rover that tests only the essentials.

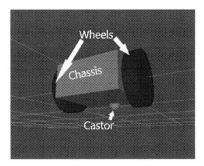

Figure 4-2. *Simplified virtual rover we are going to build*

We will build the simplified virtual rover shown in Figure 4-2 using RViz, a 3D modeling tool for ROS. RViz designs and simulates 3D components, such as wheels and sensors. Besides defining the dimensions of the components (*HxWxD*), we can model characteristics (color, material, etc.) and behavior (speed, intelligence, etc.). RViz can display 2D and 3D data from optical cameras, IR sensors, stereo cameras, lasers, radar, and LiDAR. RViz lets us build and test individual components and systems. It also offers limited testing of component interactions in the environment. Finally, RViz tests both the virtual and physical rover. Thus, we can catch design and logic errors in the simulator before and after building the hardware. We can debug the AI rover's sub-system nodes and routines inexpensively using RViz.

To start ROS communicating with the Rviz, we will open three terminal windows (Figure 4-3) using Terminator. In terminal 1, we start the master node with the `roscore` command (orange). In terminal 2, we start the RViz program with `rosrun rviz rviz` (red). The `rosrun` command takes two arguments: the ROS package the script is located in (`rviz`) and the script to run (`rviz`). The program RViz will "pop up" on your screen. Minimize it to run the final command.

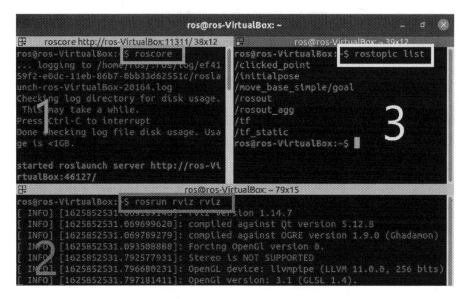

Figure 4-3. *Terminator displaying three terminals*

Finally, in terminal 3, we verify that roscore is communicating with rviz by running the rostopic list (yellow). The output shown lists the active pipelines between the nodes running in ROS—those in the yellow boxes belong to rviz and roscore. A pipeline is a computer science term that describes the dedicated pathway for passing messages between components. We will be using these pipelines later on, along with rostopic, to look at the messages being passed.

After clicking on the RViz quick launch icon, the RViz program runs by executing the rosrun rviz rviz command, as in Figure 4-4.

Note   If the rosrun rviz rviz command generates an error message, verify the line ~/catkin_ws/devel/setup.bash is in your .bashrc file in your home directory.

If the rosrun rviz rviz command still does not work, then reinstall the entire ros-noetic-desktop-full installation package. Examine the printout and determine if there are any installation errors following the `ros-noetic-desktop-full` reinstall.

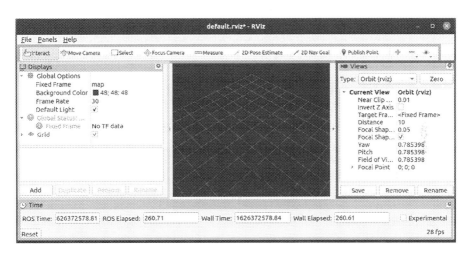

Figure 4-4. *RViz user interface*

There are four default panels in Figure 4-4: **Tools** (orange); **Views** (blue); **Displays** (yellow); and **Time** (green). We will ignore the **Time** panel since we will not use it. The center window is not a panel; it is the visualization of our virtual world. It is currently empty, with a grid placeholder. Assuming you understand **Files** and **Help** menu items, the only interesting menu item is **Panels**. The **Panels** menu item opens and closes different "panels." Panels are different ways of interacting with the current model. We will explain the different panels in more depth as needed. The Tools panel lets us work/experiment with objects in the Views panel:

- **Interact:** Reveal interactive markers.

- **Move Camera:** Move the camera around in the **Views** panel with the mouse or keyboard.

- **Select:** Point-and-drag a wireframe box around the 3D objects.

- **Focus Camera:** Focus on a point or an object.

- **Measure:** Measure distances between objects.

- **2D Pose Estimate:** Determine or plan the distance the rover traverses.

- **Publish Point (Not Seen):** Publishes coordinates of an object.

Note   Rviz tutorials can be found at the following locations:

`http://wiki.ros.org/RViz/Tutorials` and `http://wiki.ros.org/RViz/UserGuide`

The **Displays** panel interactively adds, removes, and renames the interactions of components of objects modeled in the **virtual world** (Figure 4-4). In other words, when you create a rover chassis it is modeled as a RobotModel. The **Displays** panel allows you to display the chassis axis, speed vector, etc. for a given object. The **Add** button presents appropriate graphical elements for your modeled object (in this case, the default grid), such as Camera, PointCloud, RobotModel, Axes, and Map. Selecting an element provided in the Display panel will show the description box (Figure 4-5). If you select *OK*, a 3D axis will display the default grid to show its orientation in the virtual world.

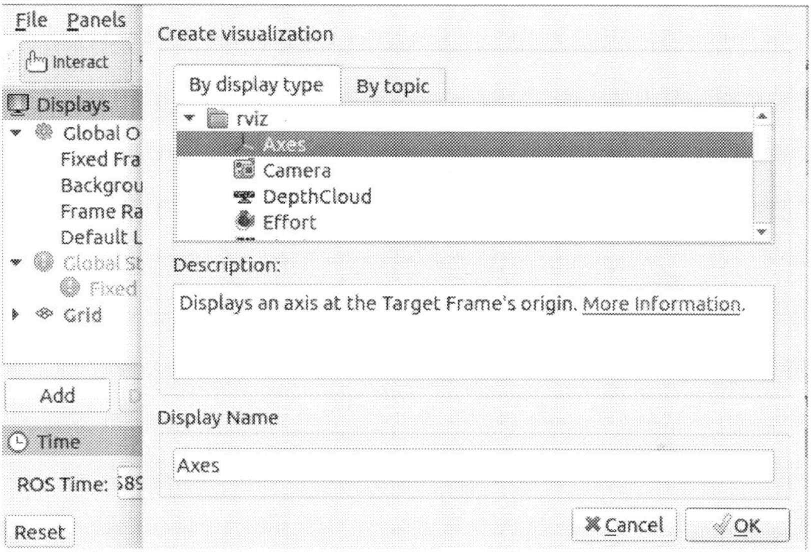

Figure 4-5. *Create visualization options by display types and by topic*

The right side of the RViz graphical user interface (GUI) layout is concerned with the **Views** panel (Figure 4-6). It controls which camera we are using to view the virtual world. The default is **Orbit**, which simulates a camera in "orbit" around our world. The other two cameras we might use are the **FPS** (first-person shooter) and **ThirdPersonFollower**. These are "gaming" terms. To understand these terms, imagine a scene of a murder. The scene has a perpetrator (first-person shooter), a victim (second person), and a witness (third person). So the **FPS** camera is from the object's eyes, while the **ThirdPersonFollower** is from the perspective of someone "witnessing" the actions (a third person).

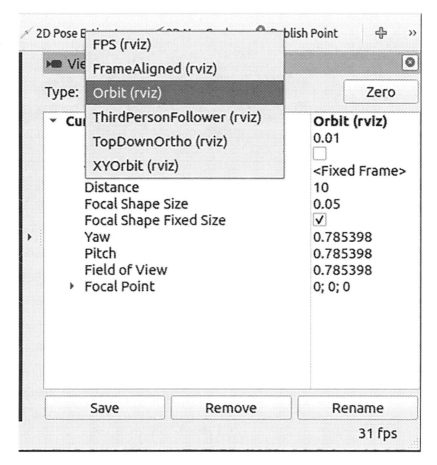

Figure 4-6. *View options and time displays*

Catkin Workspace Revisited

Recall creating a Catkin workspace in Chapter 3 for our quick testing of RViz and Gazebo. It was six steps, so we will add six steps to organize our project, as follows:

1. `cd ~/catkin_ws/src`

2. `catkin_create_pkg ai_rover ' new line`

3. `cd ~/catkin_ws`

4. `mkdir src/ai_rover/urdf ' new line`

5. `mkdir src/ai_rover/launch ' new line`

6. `catkin_make`

After executing these six commands, you will have the following directory structure (Figure 4-7). The important folder names are in bold, and a description of each folder is at the bottom of the box. (Folders we will not be explicitly using are left off Figure 4-7 for simplicity.) The fourth step executes the catkin_make file generated during the second step, catkin_create_pkg. The catkin_make script generates the other folders and related files.

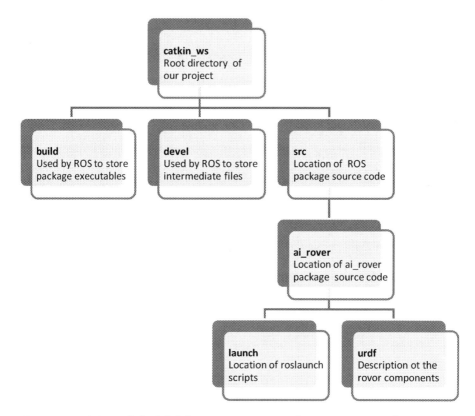

Figure 4-7. *Simplified folder organization for AI rover package*

This is the "required" folder structure for ROS projects. The root directory is catkin_ws and is hardcoded in the catkin_make script. The build and devel directories contain libraries and scripts needed to compile and execute projects. When developing ROS scripts applicable to all packages, we store the files in the src directory. The ai_rover (sub-)directory contains scripts specific to the AI rover project. The URDF directory contains the description of the rover components. There are two other files of interest: **CMakeLists.txt** and **package.xml. DO NOT EDIT!** CMakeLists.txt is created in two folders, src and ai_rover, and is used to compile scripts in their respective folders. The other file, package.xml, sets up the XML system for the AI rover.

The Relationship Between URDF and SDF

The Universal Robot Description Format (URDF) file describes the logical structure of the AI rover. The Rviz-readable URDF file is formatted in Extensible Markup Language (XML), which is a set of rules for encoding objects in a human-readable format. The URDF file contains the static dimensions of all the environment objects, such as walls, obstructions, and the AI rover and its components, along with any dynamic parameters used by those objects. The UDRF is the description (model) of the initial state of the AI rover and the environment. However, to dynamically run our AI rover in Gazebo, we need to convert the URDF file to a Simulation Description File (SDF) using GZSDF (Figure 4-8).

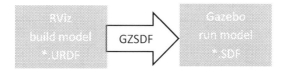

Figure 4-8. *Relationship between RViz and Gazebo in ROS development*

The SDF file uses the initial static, dynamic, and kinematic characteristics of the AI rover described in the URDF to initialize the animated AI rover in Gazebo. For example, sensor-, surface-, texture-, and joint-friction properties all can be defined within the URDF file and converted into an SDF file. We can define dynamic effects in the URDF file that might be found within the environment, such as cave-ins, collapsing floors, and explosions caused by methane build-ups. Whenever you want to add a component to the AI rover, you put it into the URDF and then convert it to the SDF.

Building the Chassis

Two required components need to be modeled in each URDF file. The *link component* is responsible for describing the static physical dimensions, orientation, and material of each object. The *joint component* describes dynamic physics, such as the amount of friction and rotational characteristics between objects.

The chassis of our AI rover is a very simple 3D box. (Download the source code at Apress: `https://github.com/Apress/Intelligent-Autonomous-Drones-with-Cognitive-Deep-Learning`) Change to the URDF directory, by entering the following terminal commands (cd `~/catkin_ws/src/ai_rover/urdf`). Create the `ai_rover.urdf` file using Gedit. Type the following terminal commands and enter the code:

```xml
<?xml version='1.0'?>
<robot name="ai_rover">
    <!-- Base Link -->
    <link name="base_link">
        <visual>
            <origin xyz ="0 0 0" rpy="0 0 0" />
                <geometry>
                    <box size="0.5 0.5 0.25"/>
                </geometry>
        </visual>
    </link>
</robot>
```

This describes our **chassis** as a 3D box 0.5 m long, 0.5 m wide, and 0.25 m tall located at the origin (0,0,0) with no rotation (no roll, no pitch, no yaw). (Most simulators use the metric system.) The chassis' `base_link` is the **link component.** All other link components will be defined relative to this `base_link`. Constructing the rover is similar to building a robot in real life; we will add pieces to the chassis to customize our rover. We use this initial `base_link` of the chassis to define the AI rover's initial position.

Using the ROSLAUNCH Command

The roslaunch command is used to launch external programs within the ROS environment, such as RViz and Gazebo. We use the roslaunch command to display the URDF files located in the URDF directory in RViz. The roslaunch command automatically starts the roscore master node for every ROS session. The roslaunch configuration file has the .launch file extension and must be located in the launch directory. In the launch directory, create the configuration file using gedit RViz.launch at the command prompt. Add the following lines:

```
<launch>
  <!-- values passed by command line input -->
  <arg name="model" />
  <arg name="gui" default="False" />

  <!-- set these parameters on Parameter Server -->
  <param name="robot_description"
  textfile="$(find ai_rover)/urdf/ai_rover.urdf" />$

  <param name="use_gui" value="$(arg gui)" />

  <!-- Start 3 nodes: joint_state_publisher,
          robot_state_publisher and rviz -->

  <node name="joint_state_publisher"
    pkg="joint_state_publisher"
    type="joint_state_publisher" />

  <node name="robot_state_publisher"
    pkg="robot_state_publisher"
    type="state_publisher" />

  <node name="rviz" pkg="rviz" type="rviz"
    args="-d $(find ai_rover)/urdf.rviz"
```

```
        required="true" />
</launch>
```

The roslaunch file has the following sections:

- Import the ai_rover.urdf model.

- Start the joint_state_publisher, robot_state_
  publisher, and the RViz 3D CAD environment.

Note   ALL URDF/SDF files must be executable:

$ sudo chmod +rwx RViz.launch

The general format of the roslaunch command is: roslaunch
<package_name> <file.launch> <opt_args>, where package_name is the
package, file.name is the configuration file, and opt_args are optional
arguments needed by the configuration file. To launch our simple chassis,
the command is as follows:

$ roslaunch ai_rover RViz.launch model:=ai_rover.urdf

Interpreting this command, "Launch RViz with the ai_rover package
using the RViz.launch configuration file, which in turn will use the ai_rover.
urdf as the model to run." The RViz screen should look like Figure 4-9. The
small red box in the middle is our "chassis."

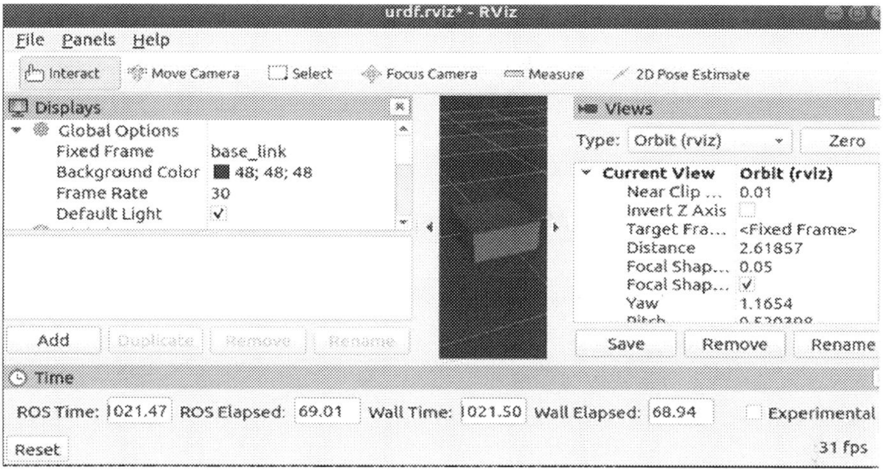

Figure 4-9. *Simple Rover chassis*

If there is no 3D box, examine the **Displays** panel to determine if **RobotModel** and **TF** (model transform) are defined. If not, do the following:

- Select the **Add** button and add **RobotModel**.

- Select the **Add** button and add **TF**.

- Finally, go to the **Global Options ➤ Fixed Frame** option and change the value to base_link.

There should now be a box on the main screen. Save your work!

Creating Wheels and Drives

Next, add the 3D links for the wheels and drives to our model. Remember, in ROS, a link is the "physical" structure between "joints." Joints are where the movement happens. Think of a human skeleton: the shoulder and elbow joints are linked by the radius bone. There are six joint types, which are defined by degrees of freedom (DoF) about the XYZ axis:

- **Planar Joint:** This joint allows movement in a plane. An example of this would be an elbow joint. (one DoF: rotate)

- **Floating Joint:** This type of joint allows motion in all six DoF (translate, rotate for each axis). An example of a joint such as this would be a wrist.

- **Prismatic Joint:** This joint slides along an axis and has a limited upper and lower range of distance to travel. An example of this would be a spyglass telescope. Think pirate telescope. (two DoF: translate and rotate)

- **Continuous Joint:** This joint rotates around the axis like the wheels of a car and has no upper or lower limits. (one DoF: rotate)

- **Revolute Joint:** This joint rotates around an axis, similar to continuous, but has upper and lower bounds of angles of rotation. For example, a volume knob. (one DoF: rotate)

- **Fixed Joint:** This joint cannot move at all. All degrees of freedom are locked. An example would be the static location of a mirror on a car door. (zero DoF)

We need to attach wheels to our chassis, and the correct joint to use is the continuous joint, because wheels rotate 360° continuously. Each wheel can rotate in both forward and reverse directions. To add the wheels to our model, modify the `ai_rover.urdf` file by adding the bold lines. Save the file after making the edits.

```
<?XML version='1.0'?>
<robot name="ai_rover">

    <!-- Base Link -->
```

```
<link name="base_link">

    <visual>
        <origin xyz="0 0 0" rpy="0 0 0" />
        <geometry>
            <box size="0.5 0.5 0.25"/>
        </geometry>
    </visual>
</link>

<!-- Right Wheel -->
<link name="right_wheel">
    <visual>
        <origin xyz="0 0 0" rpy="1.570795 0 0" />
        <geometry>
            <cylinder length="0.1" radius="0.2" />
        </geometry>
    </visual>
</link>

<joint name="joint_right_wheel" type="continuous">
    <parent link="base_link"/>
    <child link="right_wheel"/>
    <origin xyz="0 -0.30 0" rpy="0 0 0" />
    <axis xyz="0 1 0"/>
</joint>

<!-- Left Wheel -->
<link name="left_wheel">
    <visual>
        <origin xyz="0 0 0" rpy="1.570795 0 0" />
        <geometry>
            <cylinder length="0.1" radius="0.2" />
```

```
        </geometry>
      </visual>
    </link>

    <joint name="joint_left_wheel" type="continuous">
        <parent link="base_link"/>
        <child link="left_wheel"/>
        <origin xyz="0 0.30 0" rpy="0 0 0" />
        <axis xyz="0 1 0" />
    </joint>
</robot>
```

The following are the modifications made to the ai_rover.
urdf model:

- Each wheel has two parts, the link and the joint.

- The <link> of each wheel is defined as a cylinder with
  a radius of 0.2 m and a length of 0.1 m. Each wheel
  is located at (0, ±0.3, 0) and is rotated by $\pi/2$ (1.57...)
  radians or 90 degrees about the x-axis.

- The <joint> of each wheel defines the axis of rotation
  as the y-axis and is defined by the XYZ triplet "0, 1, 0".
  The <joint> elements define the kinematic (moving)
  parts of our model, with the wheels rotating around
  the y-axis.

- The URDF file is a tree structure with the AI rover's
  chassis as the root (base_link), and each wheel's
  position is relative to the base link.

Note   Our simplified virtual model's dimensions are not the same as the physical dimensions of the physical rover. This might cause some issues with training deep learning and cognitive networks. We will discuss these issues in Chapter 12 and beyond.

Verify and launch the modified code. Your Rviz display should be similar to Figure 4-10. If you do not receive a "**Successfully Parsed**" XML message, review your file for errors, such as spelling and syntax; i.e., forgetting a ">" or using "\" instead of "/".

Note   Always test file correctness after every new component added. For example, if you add the left wheel immediately check the correctness of the XML source code within the URDF file by executing the following:

```
$ check_urdf ai_rover.urdf
```

```
$ roslaunch ai_rover ai_rover.urdf
```

These two commands (check_urdf and roslaunch ai_rover) should be executed each time the file is modified. We will use "**verify and launch**" as shorthand for these two commands.

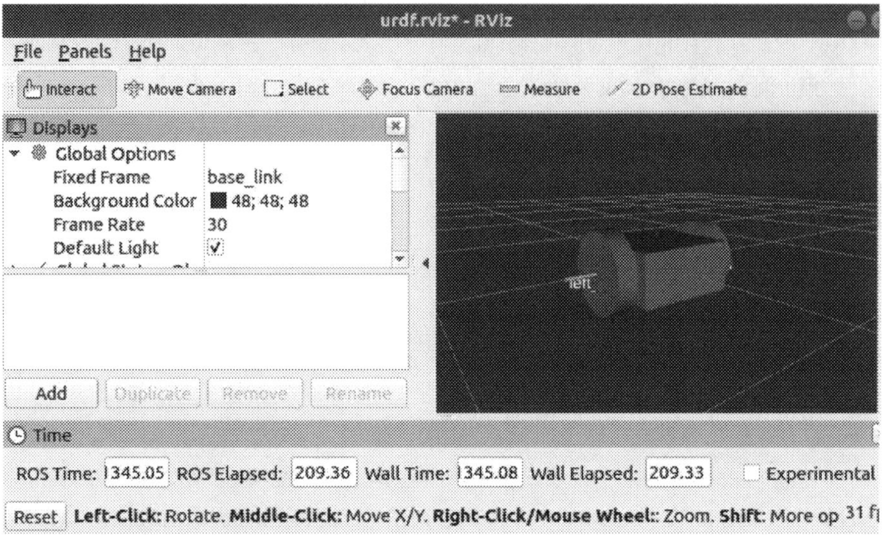

Figure 4-10. *Attaching left and right wheels to rover chassis*

Creating AI Rover's Caster

We now have the two wheels successfully attached to the AI rover's chassis. To mimic the physical GoPiGo rover, we will add a caster on the lower-back bottom of the AI rover's chassis for "balance." We could add a powered caster as a joint to add actuated turning, but this is still too complex. Instead, we will add the caster as a visual element and not as a joint. The caster slides along the ground plane as the wheels control the direction.

The highlighted changes in bold made to the `ai_rover.urdf` file add the caster as a visual element to the AI rover (`base_link`) chassis. Please note the code for the left and right wheels is collapsed, indicated by "...", and does not change.

```
<?xml version='1.0'?>
<robot name="ai_rover">
```

```
<!-- Base Link -->
<link name="base_link">

    <visual>
        <origin xyz="0 0 0" rpy="0 0 0" />
        <geometry>
            <box size="0.5 0.5 0.25"/>
        </geometry>
    </visual>

    <!-- Caster -->
    <visual name="caster">
        <origin xyz="0.2 0 -0.125" rpy="0 0 0" />
        <geometry>
            <sphere radius="0.05" />
        </geometry>
    </visual>
</link>

<!-- Right Wheel --> ...
<!-- Left Wheel --> ...
</robot>
```

We have modeled the caster as a sphere with a radius of 0.05 m (5 cm or ~2 in). After making these changes to ai_rover.urdf, **verify and launch**. Your display should look like Figure 4-11. You can see the caster sphere offset at location "0.2, 0, -0.125."

Figure 4-11. *The rover chassis with the attached caster*

Adding Color to the AI Rover (Optional)

The simple chassis modeled in `ai_rover.urdf` is constantly modified
to reflect new design requirements. For instance, to modify the color of
the chassis and wheels, we set the material color. The code in bold has a
few interesting points: 1) if you define a color (blue) in the parent link, it
affects the sub-links (`base_link/castor`); 2) if you define a color (black),
it can be reused (left/right wheel); 3) each component's `<material>` color
is located in the `<visual>` block, which must be inside a `<link>` block;
and 4) the color of the link is a "visual" component. The last point means
that the visual component will not affect any dynamic attributes; it is
decorative only.

```
<?XML version='1.0'?>
<robot name="ai_rover">

    <!-- Base Link -->
    <link name="base_link">
        <visual>
            <material name="blue">
```

```
            <color rgba="0 0.5 1 1"/>
            </material>
        </visual>

        <!-- Caster -->
        <visual name="caster">
            <origin xyz="0.2 0 -0.125" rpy="0 0 0" />
            <geometry>
                <sphere radius="0.05" />
            </geometry>
        </visual>
    </link>

    <!-- Right Wheel -->
    <link name="right_wheel">
        <visual>
            <material name="black">
                <color rgba="0.05 0.05 0.05 1"/>
            </material>
        </visual>
    </link>

    <!-- Left Wheel -->
    <link name="left_wheel">
        <visual>
            <material name="black"/>
        </visual>
    </link>
</robot>
```

At the command window, **verify and launch**. Your RViz display should look something like Figure 4-12.

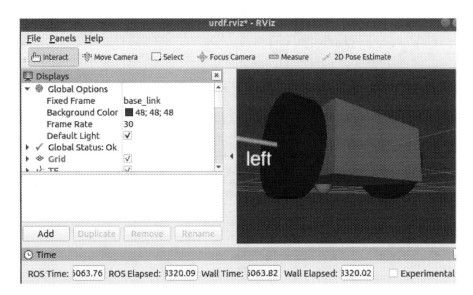

Figure 4-12. *AI rover chassis color change*

Collision Properties

Our simple model is finished enough to define the collision properties for the model—think of a collision property as a "bounding box." The bounding box is the smallest box/sphere/cylinder that surrounds our model's components, and the sum of the bounding boxes for the components is the bounding box for the rover. To do this, we add <collision> properties to each component. The collision properties are defined for Gazebo's collision-detection engine. For each simulation time frame, the components are checked for a collision. Modeling our AI rover as many simple components optimizes collision detection.

The <collision> code properties are identical to the <origin> and <geometry> properties of each component—just copy and paste between <collision>...</collision> tags. The XML source code between the <visual>...</visual> blocks was collapsed in order to save space and highlight the new <collision> blocks:

```xml
<?xml version='1.0'?>
<robot name="ai_rover">

    <!-- Base Link -->
    <link name="base_link">

        <visual>...</visual>
        <!-- Box collision -->
        <collision>
            <origin xyz="0 0 0" rpy="0 0 0" />
            <geometry>
                <box size="0.5 0.5 0.25"/>
            </geometry>
        </collision>

        <!-- Caster -->
        <visual name="caster">...</visual>

        <!-- Caster Collision -->
        <collision>
            <origin xyz="0.2 0 -0.125" rpy="0 0 0" />
            <geometry>
                <sphere radius="0.05" />
            </geometry>
        </collision>
    </link>

    <!-- Right Wheel -->
    <link name="right_wheel">
        <visual>...</visual>

        <!-- Right Wheel Collision -->
        <collision>
            <origin xyz="0 0 0" rpy="1.570795 0 0" />
```

```
      <geometry>
         <cylinder length="0.1" radius="0.2" />
      </geometry>
      </collision>
  </link>
  </joint>

  <!-- Left Wheel -->
  <link name="left_wheel">
      <visual>...</visual>

      <!-- Left Wheel Collision -->
         <collision>
              <origin xyz="0 0 0" rpy="1.57 0 0" />
            <geometry>
                <cylinder length="0.1" radius="0.2" />
            </geometry>
         </collision>
  </link>
  </joint>

</robot>
```

Verify and launch. Since the collision properties affect the dynamic physics, not the looks, you will not see any visual differences! This allows them to "bump" into other objects.

Testing the AI Rover's Wheels

Now we will test the wheels to see if they can rotate correctly. To perform these tests, we launch a GUI pop-up screen to test the wheel joints. **Verify and launch** with a small change:

```
$ check_urdf ai_rover.urdf.
```

```
$ roslaunch ai_rover ai_rover.urdf gui:=true
```

We will call this **verify and launch–GUI**. We can visualize movement!

Note   If you get a "GUI has not been installed or available" error message, run the following:

```
$ sudo apt-get install ros-noetic-joint-state-
publisher-gui
```

This forces the GUI to install.

Recall that every time that we execute the RViz.launch file, three specific ROS nodes are launched: **joint_state_publisher, robot_state_publisher,** and **RViz**. The joint_state_publisher node maintains a non-fixed joints list, such as the left and right wheels. Every time the left (right) wheel rotates, the joint_state_publisher sends a JointState message from the left (right) wheel to RViz to update the drawing of the left (right) wheel. Since each wheel generates its messages, the wheels rotate independently. After **verify and launch–GUI**, your display should look like Figure 4-13. Since the wheels are solid black, you cannot see the rotation, so you will need to launch the joint_state_publisher window. The window will display changes to the different wheels as they occur during simulation; set the initial values before simulation and modify values during simulation. These are very powerful debugging tools that you might use frequently.

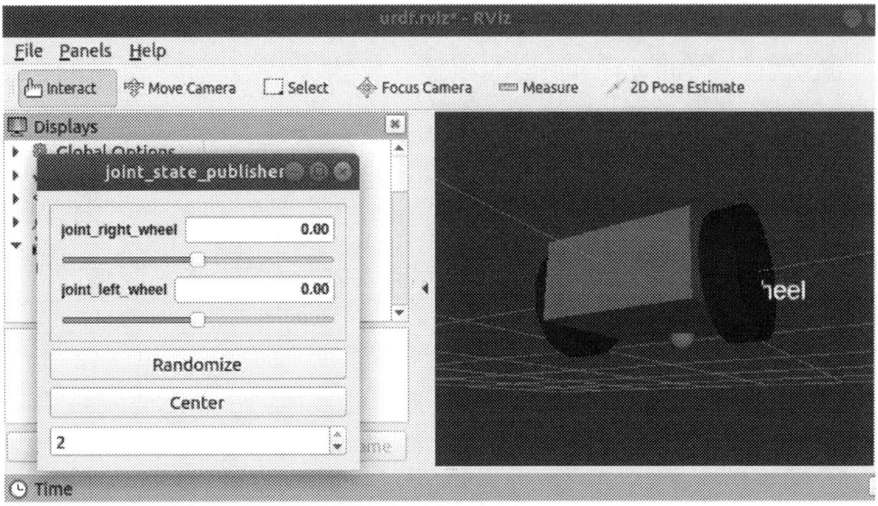

Figure 4-13. *AI rover wheel joint test GUI (joint_state_publisher)*

Examining the joint_state_publisher GUI, you should see four items of interest:

- **joint_right_wheel**: Set the angle of the wheel between ±π.

- **joint_left_wheel**: Set the angle of the wheel between ±π.

- **Randomize**: Randomly assign a value between ±π for each independent wheel.

- **Center**: Set both wheels to zero radians.

Physical Properties

Notice that our wheels are spinning, but the AI rover chassis is not moving. To see the movement, we need to do two things: add physics properties and run the AI rover in Gazebo. RViz visualizes the components but does not show the physics (movement); we need to add `inertial` properties (`mass` and `inertia`) for each component.

An object's **Inertial** is calculated from its weight and how much it resists acceleration or deceleration. For simple objects with geometric symmetry, such as a cube, cylinder, or sphere, the moment of inertia is easy to calculate. Because we modeled the AI rover with simple components, Gazebo's optimized physics engine quickly calculates the moment of inertia.

This means the chassis, wheels, and caster will all have a unique mass and inertia. Every `<Link>` element being simulated will also need an `<inertial>` tag. The two sub-elements of the inertial element are defined as follows:

> `<inertial>`
>
> > `<mass>`: Weight of the object measured in kilograms.
> >
> > `<inertia>`: The frame of a 3X3 rotational inertia matrix. The moment of inertia is defined for 3D space.
>
> `</inertial>`

Since the inertia is reflective (x ➜ z is the same as z ➜ x), we only need six elements of the matrix to fully define the moment of inertia. Each component (chassis, wheels, and caster) must have the six-element `<inertia>` values defined, as highlighted in bold.

IXX	Ixy	Ixz
Ixy	IYY	Iyz
Ixz	Iyz	IZZ

Updating the ai_rover.urdf file with the <inertial> properties for each component gives Gazebo enough information to calculate the <mass> and <inertia> for the entire rover. The modifications to the source code are highlighted in bold:

```xml
<?xml version='1.0'?>
<robot name="ai_rover">

    <!-- Base Link -->
    <link name="base_link">

        <visual>   </visual> ....
        <!-- Box collision -->
        <collision>   </collision>

        <inertial>
            <mass value="5"/>
            <inertia ixx="0.13" ixy="0.0" ixz="0.0"
                iyy="0.21" iyz="0.0" izz="0.13"/>
        </inertial>

        <!-- Caster -->
        <visual name="caster">...</visual>
        <!-- Caster Collision -->
            <collision>...</collision>

        <inertial>
            <mass value="0.5"/>
            <inertia ixx="0.0001" ixy="0.0" ixz="0.0"
```

```
        iyy="0.0001" iyz="0.0" izz="0.0001"/>
      </inertial>
  </link>

  <!-- Right Wheel -->
  <link name="right_wheel">
    <inertial>
      <mass value="0.5"/>
      <inertia ixx="0.01" ixy="0.0" ixz="0.0"
          iyy="0.005" iyz="0.0" izz="0.005"/>
    </inertial>
  </link>

  <joint name="joint_right_wheel"
type="continuous">...</joint>

    <!-- Left Wheel -->
    <link name="left_wheel">
        <inertial>
            <mass value="0.5"/>
            <inertia ixx="0.01" ixy="0.0" ixz="0.0"
                iyy="0.005" iyz="0.0" izz="0.005"/>
        </inertial>
    </link>

    <joint name="joint_left_wheel" type="continuous"> ...</joint>
</robot>
```

Each component has been defined with its unique mass and moment of inertia values. **Verify and launch–GUI**! We should see the same display in RViz and the GUI tester (Figure 4-13).

Gazebo Introduction

The UML component diagram in Figure 4-14 describes the structure of the static, dynamic, and environmental libraries of the RViz and Gazebo programs. This is why we break up our very complex problem the way we do. These are not "classes," but rather a higher abstraction that helps us organize the "libraries of components" we will need to solve our problem.

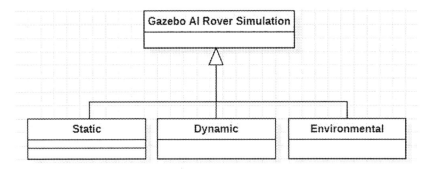

Figure 4-14. *Object generalization in Gazebo simulator*

The URDF file describes the static (color, size, etc.) and dynamic (inertial) properties of the components. Convert the URDF file into the Simulation Description Format (SDF) file for Gazebo.

We can now import the AI rover model into the Gazebo simulator. We tested the AI rover in the Gazebo physics engine to be certain everything developed is syntactically correct (check_urdf) and operational (roslaunch). Now, we integrate a simulated differential-drive motor and controller. These sensors are the beginnings of the AI rover's autonomous navigation. To model the internal mechanisms of our AI rover, we use the **urdf_to_graphiz** tool. Force the installation of urdf_to_graphiz with:

```
$ sudo apt-get install liburdfdom-tools.
```

The urdf_to_graphiz tool generates a PDF file with a logical model of the AI rover hardware (Figure 4-15). The graphical information from the tool organizes the hardware design of the AI rover. The diagram helps us visually understand relationships among the rover components. Figure 4-15 illustrates the hardware relationship from our current ai_rover.urdf model to component geometries. To display the visual model in Figure 4-15, execute the following lines (evince is a PDF reader):

```
$ urdf_to_graphiz ai_rover.urdf
$ evince ai_rover.pdf
```

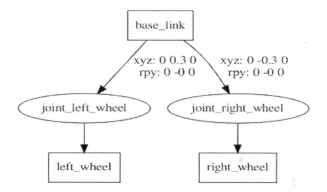

Figure 4-15. *AI rover joint wheel connections*

Background Information on Gazebo

We will be using the Gazebo simulator for AI rover experimentation. The simulator supplies multiple development and deployment utilities. Typical Gazebo applications are the following:

- Development of deep learning algorithms

- Development of control algorithms

- Simulation of sensor data for LiDAR systems, cameras, contact sensors, proximity sensors, etc.

- Advanced physics engines via open dynamics engine

Now we are reviewing the actual process of loading the URDF description of the AI rover into Gazebo. We will first test the AI rover model by taking control of the wheels to move, in a limited fashion, the AI rover model within a simulated world with obstacles. This will be done at first without the use of a two-wheeled differential-drive control system. We will develop that later, in the advanced sections of this chapter, by extending our AI rover model to have the independent ability to control its very own continuous wheel joints, graph sensor data, and verify and validate control and deep learning algorithms.

Starting Gazebo

To test whether Gazebo has been installed correctly, we can enter the following Linux terminal command:

```
$ Gazebo.
```

If Gazebo is not installed, refer to Chapter 3.

Every time that Gazebo is run, two different processes are created. The first is the Gazebo Server (`gzserver`), which is responsible for the overall simulation. The second process is the Gazebo Client (`gzclient`), which starts the USER GUI used to control the AI rover.

Note   If you execute the $ Gazebo Linux terminal command and get a series of errors or warning messages, you may have previous incarnations of ROS nodes running. Execute the $ rosnode list command to determine if there are any previously running nodes.

If there are any ROS nodes still active, simply execute $ `rosnode kill -a`. This command kills all running ROS nodes. Then, simply run the $ `Gazebo` command once again. Be certain to always check for any node warning messages.

A successful launch of Gazebo will create a window similar to Figure 4-16.

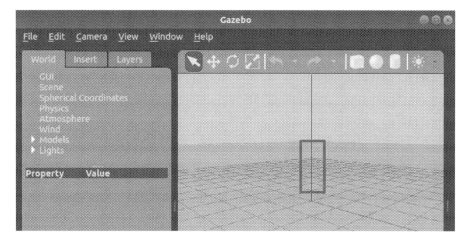

Figure 4-16. *Gazebo screen*

There are two main areas: the simulation display window and the tabs panel. The simulation display window is where our generated world (and rover) will be displayed. The toolbar that is located at the very top of the simulation display window symbols controls the simulated world. (Note the little red box; we will come back to this a moment.) The tabs panel has three tabs: World, Insert, and Layers.

The World tab provides hierarchical access to sub-elements, such as **GUI, Scene, Spherical Coordinates, Physics, Models,** and **Lights.** While all of these categories are fascinating, at this time we are interested in the Models tab—where our AI rover model resides. We will introduce other categories as needed.

The Insert tab gives access to models developed by us (local) and others (cloud, located at `http://gazebosim.org/models`). These models may be inserted into our active world.

The Layers tab allows toggling between different visual parts of our simulated world. We use this to "debug" our world view; for instance, determining if there are any unexpected collisions. The Layers tab initially contains no layers. As we develop our world further, we can add layers.

Gazebo Environment Toolbar

The toolbar is located at the very top of the Gazebo environment. Let's review the following symbols that can be seen from left to right within the Gazebo toolbar. These symbols have the following capabilities and can also be seen in Figure 4-17.

The following symbols can be seen from left to right within the Gazebo toolbar. See Figure 4-17.

Figure 4-17. *The Gazebo environment toolbar*

- **Selection Mode:** This mode selects the 3D AI rover or its components within the Gazebo environment. The properties of the AI rover or its components are listed within the World panel.

- **Translation Mode:** This mode selects the AI rover or its components when a cursor is clicked around any part of the AI rover. There will be a 3D box wrapped around the selected component or even the AI rover itself. We can then move any part of the AI rover to any position required.

- **Rotation Mode:** This mode is responsible for selecting the AI rover model when a cursor selects and draws a box around it. You can then rotate the AI rover model on either its roll, pitch, or yaw axis.

- **Scale Mode:** This mode can select the AI rover sub-components, such as the box component. The scaling operation only works with very simple 3D shapes, such as a cube in the case of the chassis for the AI rover.

- **Undo Command:** This will undo the very last action committed by the developer. We can repeat the undo operation to undo a series of actions in a linear format.

- **Redo Command:** This likewise will redo the last action that was deleted by the undo command. So it will reverse and restore what was eliminated by the undo command.

- **Box, Sphere, and Cylinder Modes:** These next three modes found by their shapes allow one to automatically create these shapes with varying dimensions within the Gazebo environment. The scaling mode can be used to modify the dimensions of these simple shapes.

- **Lighting Mode:** This allows one to change the angle and intensity of light within the Gazebo environment.

- **Copy Mode:** Copies the selected items within the Gazebo environment.

- **Paste Mode:** This mode pastes the copied item onto the Gazebo environment.

- **Selection and Alignment Mode:** This mode will align two objects with each other in either the x, y, or z-axis.

- **Join Mode:** This mode will allow one to select the location as to where two objects will be joined.

- **Alter View Angle Mode:** This mode will allow one to change the angle of view for the user.

- **Screenshot Mode:** This mode will take a screenshot of the simulation environment for documentation purposes. All files are saved within the `~/gazebo/pictures` directory.

- **Log Mode:** This information will take all of the data and simulation values being generated and store them in the `~/gazebo/log` directory. This will be used to debug the deep learning routines for the AI rover.

The Invisible Joints Panel

We now revisit the red box in Figure 4-16. Dragging the dotted control to the left accesses the Joints panel—our testing interface for any active model; for instance, our rover. Dragging the control will then create a display similar to Figure 4-18.

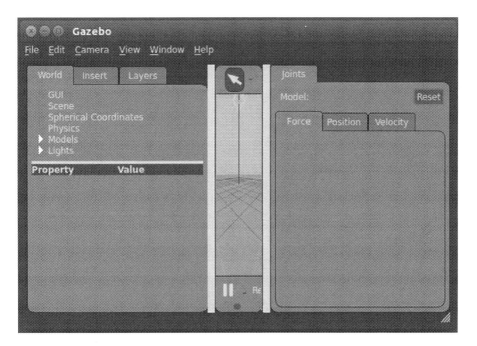

Figure 4-18. *Gazebo Joints panel screen pulled from right to left*

The Joints panel has one *Reset* button and multiple tabs. The *Reset* button will return our active model to its initial configuration. The tabs display the active model's available joints and properties. In our AI rover, the only available joints will be the two wheels. The three tabs are defined as follows:

- **Force:** Defined as a force in Newtons per meter (N/m) applied to each continuous joint.

- **Position:** <x,y,z> 3D coordinates and <roll,pitch,yaw> rotation.

- **Velocity:** Speed of Joint in meters per second (m/s). These can also be set by the PID values.

The Gazebo Main Control Toolbar

Now that we have reviewed the toolbar that is used to control the shapes, dimensions, and operations that occur within a Gazebo simulation, we must review the basic controls of the controlling toolbar that is located at the far-most-top of the Gazebo environment itself. This basic controlling toolbar contains the **File, Edit, Camera, View, Window,** and **Help** options that would be expected to be found in any modern GUI environment. We will now review each of the following basic functionalities found within the controlling toolbar of the Gazebo environment:

- **File** has the sub-functions of Save World, Save World As, Save Configuration, Clone World, and Quit.

- **Edit** has the sub-functions of Reset Model Poses, Reset World, Building Editor, and Gazebo Model Editor.

- **Camera** has the sub-functions of Orthographic, Perspective, FPS View Control, Orbit View Control, and Reset View Angle.

- **View** has the sub-functions of Grid, Origin, Transparent, Wireframe, Collisions, Joints, Center of Mass, Inertias, Contacts, and Link Frames.

- **Window** has the sub-functions of Topic Visualization, Oculus Rift Virtual Reality Viewer, Show GUI Overlays, Show Toolbars, and Full Screen.

- **Help** has the sub-functions of Hot Key Chart and Gazebo About.

Now that we have reviewed the controlling toolbar functions, we must transition how we can run simulations and play back simulation runs. We must be able to modify the URDF file of the AI rover into a form that is compatible with the Gazebo simulation environment by transforming that very same URDF file into an SDF (Simulation Defined Format) file.

URDF Transformation to SDF Gazebo

Now we must transform the AI rover's URDF file so that it can be accepted and processed by the Gazebo environment. We must convert the URDF to an SDF file. We must state to the reader that the SDF expression is an extension to that of URDF, by using the same XML extensions provided. By making the appropriate modifications of the URDF file describing the AI rover, it will allow the Gazebo environment to convert the URDF to the required SDF robot expression for the AI rover. We will now describe the required steps to transform URDF files into SDF files.

To allow this transformation to be complete, we must add the correct <Gazebo> tags to the URDF file that describes the AI rover chassis, wheels, and caster within the Gazebo simulator. It should be stated that the chassis of the AI rover not only includes the physical box of the AI rover but would also include the mass and moment of inertia of the embedded electronics, such as the Raspberry Pi. The use of the <Gazebo> tag allows one to transform elements found in SDF but not in URDF. If a <Gazebo> tag is used without a reference="" property, then the <Gazebo> tag is concerned with the entire AI rover model. The reference parameter usually refers to joints such as the wheels defined within the AI rover URDF file. We can also define both links and joints found within SDF that are not found within the URDF file describing the AI rover. With this extension found within SDF, we can develop sophisticated simulations of a deep learning controller controlling the AI rover within a Gazebo environment. We will review in this and the next chapter some of the tutorials found within http://gazebosim.org/tutorials/?tut=ros_urdf for a list of elements such as links and joints

149

that can be used to even further enhance the simulations of the AI rover. Examples of links and joints would be the fixed sensors and dynamic actuators for the AI rover.

Not only can we define the links and joints within Gazebo, but we can also define and specify the color within Gazebo. However, we have to make modifications in Gazebo that are different than the AI rover model definitions that were defined within Rviz. For example, we cannot reuse the references defined for the color of the components. As such, we must add a Gazebo <material> for each link. The Gazebo tags can be placed in the AI rover model before the ending </robot> tag, as follows:

```
<gazebo reference="base_link">
    <material>Gazebo/Blue</material>
</gazebo>

<gazebo reference="right_wheel">
    <material>Gazebo/Black</material>
</gazebo>

<gazebo reference="left_wheel">
    <material>Gazebo/Black</material>
</gazebo>
```

Gazebo tags would have to be defined before the ending </robot> tag for the entire model for the AI rover. Therefore, all Gazebo tags should be defined near the end of the file before the ending </robot> tag. However, there are caveats with the other elements in Gazebo.

The Gazebo simulator will use neither the <visual> nor the <collision> elements if they are not specified for each link, such as the AI rover 3D chassis box or the caster. If links such as these are not specified, then Gazebo will regard them as being invisible to sensors such as lasers and simulated environment collision checking.

Checking the URDF Transformation to SDF Gazebo

Just like we had to verify and validate the URDF files earlier in this chapter with the check_urdf tool found within Noetic ROS, we will also have to re-examine the URDF files that have been upgraded with the <Gazebo> extension tags. We need to do this to determine if any errors within the URDF files have the <Gazebo> extension tags that indeed need to be transformed into the SDF files required for exporting the AI rover model to the Gazebo simulation environment. We will now extend the AI_Rover. urdf file with the name extension of AI_Rover_Gazebo.urdf. The name extension is to designate that this file is to be used for Gazebo simulations for the AI rover. The designated tool that is used to verify the URDF <Gazebo> extensions that allow the URDF file to be transformed into an SDF file by and for Gazebo is the $ gz sdf toolset. Two commands are needed:

```
$ gz sdf -p ai_rover.gazebo
```

Or for an entire directory, search for gazebo extension files:

```
$ gz sdf -p $(rospack find ai_robotics)/urdf/ai_rover.gazebo
```

We will first test to determine if the Gazebo references work for the color schemes for the chassis and wheels. We will also use Gazebo references to develop the differential drive controller for the AI rover itself by this chapter's end.

Now that we've reviewed the basics of utilizing the verification process for a URDF file that has `<gazebo>` extension tags, these same extension tags must be placed between the `<link>` and `<joint>` tags of each component, such as the `base_link` and both wheels for the AI rover. We must review an example of our AI rover URDF file that has the `<gazebo>` extension tags. The examples are highlighted in bold as follows:

```
<?XML version='1.0'?>
<robot name="ai_rover">

    <!-- Base Link -->
    <link name="base_link">
    </link>

        <gazebo reference="base_link">
                <material>Gazebo/Blue</material>
        </gazebo>

    <!-- Right Wheel -->
    <link name="right_wheel">
    </link>

    <gazebo reference="right_wheel">
        <material>Gazebo/Black</material>
    </gazebo>

    <joint name="joint_right_wheel" type="continuous">
    </joint>

    <!-- Left Wheel -->
    <link name="left_wheel">
    </link>

    <gazebo reference="left_wheel">
        <material>Gazebo/Black</material>
    </gazebo>
```

```
<joint name="joint_left_wheel" type="continuous">
</joint>
```
```
</robot>
```

Once we have this URDF file with the first Gazebo extensions created, we must then convert this to an SDF file, to be certain that there are no issues with the transformation process for processing by Gazebo. We will then execute the following command: **$ gz sdf -p ai_rover Gazebo.** Once we execute this command in the correct directory, we should have a terminal prompt listing of the correct and equivalent SDF file being generated with no printed errors. Once we have reached this point of generating an SDF file, we must now develop the required launch and simulation files for starting our initial ROS simulation within Gazebo.

First Controlled AI Rover Simulation in Gazebo

As we are developing our first controlled AI rover simulation within Gazebo, we must develop two files that separate two steps for creating this simulation environment. We must first develop the launch file to launch and view the AI rover, the environment, and any obstacles or mazes presented within the simulated environment. The second file will describe what the Gazebo simulation world will contain, such as mazes, obstacles, and dangers. We should note that the second Gazebo simulation file will likewise be launched by the first launch file developed. We should also be aware that the launch file should be located within the `launch` directory and the Gazebo obstacle simulation file should be located within the `worlds` directory, all of which are sub-directories to the `ai_robotics` directory.

Therefore, this launch file will be launching the empty world as follows:

```
<launch>
  <!-- We use ROSLAUNCH AND empty_world.launch, -->
  <include file="$(find gazebo_ros)/launch/empty_world.launch">
    <arg name="world_name" value="$(find  ai_robotics)/worlds/
    ai_rover.world"/>

    <arg name="paused" default="false"/>
    <arg name="use_sim_time" default="true"/>
    <arg name="gui" default="true"/>
    <arg name="headless" default="false"/>
    <arg name="debug" default="false"/>

  </include>

  <!-- Spawn ai_rover into Gazebo -->
<node name="spawn_urdf" pkg="gazebo_ros" type="spawn_model"
output="screen"
    args="-file $(find ai_robotics)/urdf/ai_rover.gazebo
    -urdf -model ai_rover"/>

</launch>
```

We will now need to create the sub-directory worlds for the rover. This can be done with the following terminal commands:

```
$ cd ~/catkin_ws/src/ai_robotics
$ mkdir worlds
$ cd worlds
```

This launch file will launch the empty worlds that are contained within the gazebo_ros package. We can also develop a world that will contain the Egyptian catacomb layout by replacing the ai_rover.world file. The URDF with the <gazebo> extension tags model of the AI rover will be launched within the empty worlds by the spawn_model service from the gazebo_ros Noetic ROS node.

Now that we have the worlds directory created, we can begin to develop the SDF that will become the ai_rover.world file that is launched with the aforementioned launch file and includes additional items, such as construction cone obstacles. Therefore, we will closely examine the ai_rover.world file that describes the ground plane, the light source (sun), and the two separated construction cones. The source code for the ai_rover.world is the following:

```
<?XML version="1.0"?>
<sdf version="1.4">
<world name="default">
<include>
<uri>model://ground_plane</uri>
</include>
<include>
<uri>model://sun</uri>
</include>
<include>
<uri>model://construction_cone</uri>
<name>construction_cone</name>
<pose>-3.0 0 0 0 0 0</pose>
</include>
<include>
<uri>model://construction_cone</uri>
<name>construction_cone</name>
<pose>-3.0 0 0 0 0 0</pose>
</include>
</world>
</sdf>
```

We can always modify this file to include more construction cones and other obstacles by modifying and using the <include>, <uri>, <name>, and <pose> tags. The <include> tag allows us to include an additional model, such as a construction cone. The <uri> model identifies what the model is, such as a construction cone. The <name> tag identifies the name of the obstacle. The <pose> tag represents a relative coordinate transformation between a frame and its parent. Link frames were always defined relative to a model frame, and joint frames relative to their child link frames. Now we can execute the ai_rover_gazebo.launch file by executing the following command:

```
$ roslaunch ai_robotics ai_rover_gazebo.launch
```

Once you execute this terminal command, you should have the display shown in Figure 4-19.

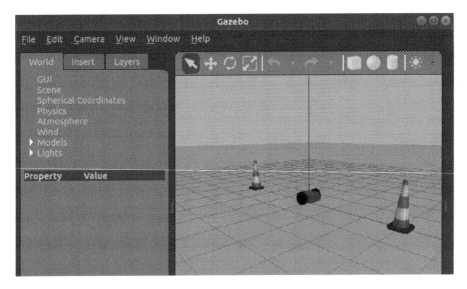

Figure 4-19. *First AI rover Gazebo simulation*

First Deep Learning Possibility

Now that we have developed the first AI rover Gazebo simulation setup, we must experiment to explore methods to cause locomotion, and then eventually intelligent navigation, obstacle avoidance, and ultimately sense-and-avoid cognitive capabilities. The first use of a deep learning controller might take the form of controlling any type of unexpected behavior of the AI rover within Gazebo. Unexpected behavior would include not traveling within a straight line while navigating obstacles. This is because the URDF file with the `<Gazebo>` extension tags might need further tuning to represent the physics within Gazebo. We might need to develop an intelligent and adaptive deep learning controller that controls the AI rover. We might need to modify properties such as the mass distribution and moment of inertia values for the AI rover. If these values were constantly changing, we would need a controller that would likewise adapt accordingly.

Moving the AI Rover with Joints Panel

Now we should try to test the underlying physics engine that is provided within Gazebo. The one effective way to accomplish this is to cause the model of the AI rover to move within Gazebo. Therefore, to test the physics engine for the AI rover we must use the joint control by using the Joints panel. The Joints panel is located to the right of the Gazebo environment. We need to be in selection mode, which we can do by clicking onto the AI rover model, which will be highlighted with a white outline box. Once the white box has been indicated, we can see the values for the `joint_left_wheel` and the `joint_right_wheel` displayed within the Force tab of the Joints panel. We need to input very small values, such as 0.00050 Newton-Meters for the `joint_left_wheel` and 0.00002 Newton-Meters for the `joint_right_wheel`. We should then see our AI rover prototype move in an arching pathway. We should try to collide the AI rover with one of the construction cones. We do this to test the collision tags found within the AI Rover URDF file to see if they work. We can see that the collision tags do indeed work by examining the crash display shown in Figure 4-20.

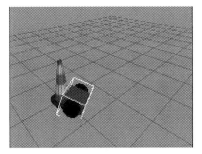

Figure 4-20. *First AI rover crash with construction cone*

Summary

We have achieved a lot within the pages of Chapter 4. We have reviewed how to develop a model for the AI rover with URDF. We have shown how to extend a URDF file with <Gazebo> tags to allow for Gazebo simulations. We have evaluated the functionality of the 3D environment for designing models such as Rviz. We have reviewed the process of developing and deploying models created in Rviz to Gazebo. We have worked with multiple ROS commands to launch these simulations. We will see in Chapter 5 how we can use the XML macro (Xacro) languages to develop even more sophisticated AI rover simulations, by allowing the AI rover, sensors, actuators, and simulated environments to be developed more efficiently. We will also use more examples of UML modeling for these very same Xacro files.

EXTRA CREDIT

Exercise 4.1: What additional changes would you make to the `ai_rover.world` file to include obstacles other than the construction cones?

Exercise 4.2: What additional changes would you make to spawn an additional number of construction cones within the `ai_rover.world`? How can you place them differently or symmetrically, etc.?

Exercise 4.3: How does the use of the Joints panel highlight the need for a controller and driver for the differential-wheeled system? Why can we not develop a differential driver within Rviz?

Exercise 4.4: Why do we need to verify and validate both the URDF and SDF files being developed with tools such as `check_urdf`?

## CHAPTER 5

Adding Sensors to Our Simulation

In Chapter 4, we created our first Gazebo simulation of the rover. We manually drove the rover around the empty environment. We used URDF files to define what our rover looked like and the launch files to define the rover controls in the Gazebo environment. Now we want to add sensors so that the rover can "see" obstacles. Unfortunately, the URDF files will get very big and hard to maintain without a little help. Enter the Xacro XML language extension (Xacro stands for XML Macro). It helps add standard sensors via simple blocks of code, called plug-ins. The plug-ins support sensors such as LiDAR, radar, cameras, etc. Our LiDAR sensor develops the first distance-ranging capabilities of the rover. We will develop Python scripts that directly control the rover's wheels through the differential drive plug-in. Finally, we will experiment with a keyboard control (Teleops) of the rover via a Python script.

Objectives

The following are the objectives required for successful completion of this chapter:

- Learn XML Macro programming language (Xacro)

- Remaster the rover model with the Xacro language for simplicity and expandability

© David Allen Blubaugh, Steven D. Harbour, Benjamin Sears, Michael J. Findler 2022
D. A. Blubaugh et al., *Intelligent Autonomous Drones with Cognitive Deep Learning*,
https://doi.org/10.1007/978-1-4842-6803-2_5

- Develop Xacro programming language routines for the sensors and motors

- Test the rover in RViz and Gazebo simulators

- Control the rover

XML Macro Programming Language

The Xacro programming language is a macro language used to develop maintainable and modular XML files. We will have a brief overview of Xacro to optimize the URDF robot and Gazebo simulation files.

A macro is a simple substitution of a "name" with a "value." The most straightforward macro is property substitution. A property is typically a constant used throughout a file. Replacing the constant with a "name" placeholder allows the programmer to define properties in a single location, allowing easier maintenance. Changing the value at that single location changes the named constant at all locations in the file. The format is as follows:

Preferred Style

```
<xacro:property name="propertyName" value="propertyValue"/>
```

Equivalent Block

```
<xacro:property>
   <name="propertyName">
   <value="propertyValue" />
</xacro:property>
```

The property name is found in an XML file, and the property value is substituted there. The value can be a simple number or a string. The following example shows how to declare and use a property:

```
<xacro:property name="myRadius" value="2.1" />
<xacro:property name="myLength" value="4.5" />

<geometry type="cylinder" radius="${myRadius}"
length="${tmyLength}" />
```

The properties substitute the geometry expression by replacing the names inside dollar-braces (${}).[1] We will use property blocks to define the dimensions of the rover chassis. If we ever want to change the size of our rover, we just change the property block.

A more sophisticated substitution allows multiple values for one property name. Here's an example of using a property block to place values for the cartesian (x, y, z) and orientation (roll, pitch, yaw) values for the geometry expression:

```
<xacro:property name="front_left_origin">
  <origin xyz="0.3 0 0" rpy="0 0 0" />
</xacro:property>

<pr2_wheel name="front_left_wheel">
  <xacro:insert_block name="front_left_origin" />
</pr2_wheel>
```

We define within the front_left_origin property block xyz values as being "0.3 0 0" and the rpy = "0 0 0". If we get a new rover with a new chassis, we could update the entire system by changing the front_left_origin property block. We do not have to track down every instance of "front_left_origin" in all the files for the required changes.

We can use Xacro for simple math expressions for sensor processing or for rover component dimensions. Only basic arithmetic and variable substitution are supported. For example:

---

[1] If you want a literal "${", you should escape it as "$${".

```
<xacro:property name="radius" value="4.3" />
<circle diameter="${2 * radius}" />

<xacro:property name="R" value="2" />
<xacro:property name="alpha" value="${30/180*pi}" />
<circle circumference="${2 * pi * R}"
    pos="${sin(alpha)} ${cos(alpha)}" />
<limit lower="${radians(-90)}" upper="${radians(90)}"
    effort="0" velocity="${radians(75)}" />
```

The Xacro ${} is also an extension for the Python arithmetic library. Both constants (**pi**) and functions (**radians**, a degrees-to-radians conversion) defined in the library are accessible.

Xacro has conditional blocks (if..unless) similar to programming languages (if..else), such as Python. The syntax format of the Xacro conditional blocks are as follows:

```
<xacro:if value="<expression>">
   <... some xml code here ...>
</xacro:if>
<xacro:unless value="<expression>">
   <... some xml code here ...>
</xacro:unless>
```

The conditional blocks must always return a Boolean value; i.e., true (1) or false (0). Any other return throws an exception, an invalid return value. In the following five statements, the first line defines var as useit. The second line checks whether var is equal to useit and returns a true. The third line looks at the substrings of var and returns a true. The fourth line defines an array of values called allowed. The fifth line checks to see if the "1" is allowed in the array. When determining the value, we used double quotes " ", and we used single quotes ' ' when using a string constant.

```
<xacro:property name="var" value="useit"/>
<xacro:if value="${var == 'useit'}"/>
<xacro:if value="${var.startswith('use') and
    var.endswith('it')}"/>
<xacro:property name="allowed" value="${[1,2,3]}"/>
<xacro:if value="${1 in allowed}"/>
```

More Examples of XML

To show the power of XML and Xacro, we will create a few example macros. This XML example declares a joint and link pair. The dynamic joint is named caster_front_left_joint with an axis value of xyz="0 0 1". The link component, caster_front_left, defines the coordinates (xyz="0 1 0"), orientation {rpy="0 0 0"), color (name="yellow"), and the mass constant (0.1). Furthermore, we define inertial (moment of inertia) values.

```
<joint name="caster_front_left_joint">
  <axis xyz="0 0 1" />
</joint>
<link name="caster_front_left">
  <pose xyz="0 1 0" rpy="0 0 0" />
  <color name="yellow" />
  <mass>0.1</mass>
  <inertial>
    <origin xyz="0 0 0.5" rpy="0 0 0"/>
    <mass value="1"/>
    <inertia ixx="100"  ixy="0"  ixz="0"
             iyy="100" iyz="0" izz="100" />
  </inertial>
</link>
```

To illustrate the capabilities of Xacro, the next macro creates a macro that receives two macros as parameters, "first" and "second"! These two macros are defined elsewhere. The two parameters inserted sequentially create a larger macro called "reorder." These multiple block parameters execute in the specified linear order. These executed parameters might pass the left/right, front/back wheels as macros in a single control macro to generate four individual control macros (left/front, right/front, left/back, right/back).

```
<xacro:macro name="reorder" params="*first *second">
  <xacro:insert_block name="first"/>
  <xacro:insert_block name="second"/>
</xacro:macro>

<reorder>
  <first/>
  <second/>
</reorder>
```

Macros may contain other macros, called nesting. The outer macro expands first, and then the inner macro expands. A full description is beyond the scope of this text. To include nested macros in the main macro, do the following:

```
<xacro:include filename="$(find Rover)/urdf/ai_rover_
remastered_plugins.xacro"/>
```

This macro searches the project directories for a filename called ai_rover_remastered_plugins.xacro and inserts that file into the current file. The ai_rover_remastered_plugins.xacro file stores the filenames of the plug-ins for the rover, such as sensor plugins (LiDAR, radar, etc.) and the two-wheeled differential drive (DDC). Built into the DDC are simple keyboard commands to control a differential drive.

The Rover Revisited

We will review and remaster the design of our two-wheeled differential-drive rover system by transitioning from standard XML to a Xacro URDF description file. The differential-drive system for the rover is the most common type of drive system for a robot and navigates by independently controlling the velocities of each of the wheels. Since the rover only utilizes two wheels and a non-moving static caster, we should consider refactoring in Xacro. The significant advantage of Xacro is that it is easier to maintain, implement, test, and expand Gazebo simulations. Xacro's modular design and Python scripting mean we can quickly test the virtual rover's routines. Xacro helps transition our designs from the Gazebo simulations to the physical rover's existing software and hardware implementation.

Recall that a differential-drive system navigates the environment by controlling each wheel's velocity independently. The left- and right-front wheels control (or actuate) navigation, and their velocities determine the rover's driving path. For additional information, please refer to `https://en.wikipedia.org/wiki/Differential_wheeled_robot`.

Modular Designed Rover

If we are going to maintain this increasingly complex software project, we should simplify its structure. At this moment, putting all the source code in one file seems to make life simple. But soon, we will be adding additional hardware and software that will make modifying the underlying changes challenging to track. So, while we are at the beginning of our design, let us simplify the packaging of our code for later. We will divide our original code into modules that have one (and exactly one) responsibility. Our original URDF code divides into the following Xacro modules:

- Dimensions (`dimensions.xacro`), which tracks the constants for the physical components of our rover

- Chassis (`chassisInertia.xacro`), which tracks the physics related to the body of our rover

- Wheels (`wheelsInertia.xacro`), which tracks the physics related to each wheel of our rover

- Caster (`casterInertia.xacro`), which tracks the physics related to the caster of our rover

- Laser (`laserDimensions.xacro`), which tracks the physics and geometry layout of the LiDAR's housing box

- Camera (`cameraDimensions.xacro`), which tracks the physics and geometry layout of the camera's housing box

- IMU (`IMUDimensions.xacro`), which tracks the physics and geometry layout of the inertial measurement unit's (IMU) housing box

After adding new sensors, we will need to modify the dimensions file and add our `<sensor>Inertia.xacro` file. Logically, our software project now looks like Figure 5-1.

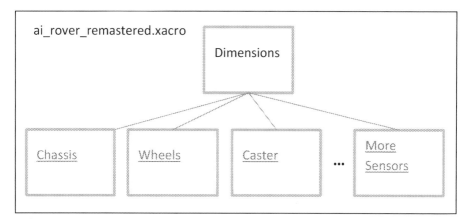

Figure 5-1. *Modular design of our rover*

To begin our code refactoring, we need to create the rover sub-directory ai_rover_remastered. The terminal commands are as follows:

```
$ mkdir -p catkin_ws/src/ai_rover_remastered
$ cd catkin_ws/src/ai_rover_remastered
$ mkdir launch urdf config
```

Now we should have three sub-directories, launch, urdf, and config, in the ai_rover_remastered directory. Recall that having the launch, urdf, and config sub-directories prepare the ai_rover_remastered directory as an ROS package. **Note that URDF supports XML and Xacro.**

dimensions.xacro

Next, create the dimensions.xacro file in the catkin_ws/src/ai_rover_remastered/urdf directory:

```
<?xml version="1.0"?>
<robot name="ai_rover_remastered" xmlns:xacro="http://www.ros.org/wiki/xacro">
<xacro:property name="base_width" value="0.16"/>
```

```
<xacro:property name="base_length" value="0.16"/>
<xacro:property name="wheel_radius" value="0.035"/>
<xacro:property name="base_wheel_gap" value="0.007"/>
<xacro:property name="wheel_separation" value="0.15"/>
<xacro:property name="wheel_joint_offset" value="0.02"/>

<xacro:property name="caster_wheel_radius" value="${wheel_
radius/2}"/>
<xacro:property name="caster_wheel_mass" value="0.001"/>
<xacro:property name="caster_wheel_joint_offset"
value="-0.052"/>
</robot>
```

We have created our first Xacro URDF file for the rover. It defines the rover's dimensions and serves as a framework for different rover models. The property block tags define numeric and string constants, such as the rover's base length, "base_length," which is statically defined as 0.16. Any Xacro files that include this file will have access to this global constant. Changing the value to 0.20 in dimensions.xacro will have a ripple effect across all files. Without Xacro, we would have to search ALL the files for "0.16," decide whether it referred to base_length or base_width, and manually change it—a very error-prone procedure. (Compare to Chapter 4's URDF.)

chassisInertia.xacro

Next, we create chassisInertia.xacro to define the movement of our rover. Notice that it "includes" the dimensions.xacro file. This chassisInertia.xacro file also includes our final ai_rover_remastered.xacro file. While this nesting of files seems complicated at first, separating the structure from the functionality allows us to change one without modifying the other.

```xml
<?xml version="1.0"?>

<robot name="ai_rover_remastered" xmlns:xacro="http://www.ros.
org/wiki/xacro">

<xacro:macro name="box_inertia" params="m w h d">
  <inertial>
    <mass value="${m}"/>
    <inertia ixx="${m / 12.0 * (d*d + h*h)}" ixy="0.0"
    ixz="0.0" iyy="${m / 12.0 * (w*w + h*h)}" iyz="0.0"
    izz="${m / 12.0 * (w*w + d*d)}"/>
  </inertial>
</xacro:macro>

<link name="base_footprint">
  <xacro:box_inertia m="20" w="0.001" h="0.001" d="0.001"/>
  <visual>
    <origin xyz="0 0 0" rpy="0 0 0" />
    <geometry>
        <box size="0.001 0.001 0.001" />
    </geometry>
  </visual>
</link>

<link name="base_link">
  <xacro:box_inertia m="10" w="${base_length}" h="${base_
  width}" d="0.01"/>
  <visual>
    <geometry>
      <box size="${base_length} ${base_width} 0.01"/>
    </geometry>
  </visual>

  <collision>
```

```
    <geometry>
      <box size="${base_length} ${base_width} 0.01"/>
    </geometry>
  </collision>
</link>

<joint name="base_link_joint" type="fixed">
  <origin xyz="0 0 ${wheel_radius + 0.005}" rpy="0 0 0" />
  <parent link="base_footprint"/>
  <child link="base_link" />
</joint>

</robot>
```

wheels.xacro

The following components are the two identical wheels. Because they are similar, we need to define the concept (class) once and instantiate (object) it twice by offsetting them by the appropriate amount specified in the generated ai_rover_remastered.urdf file.

```
<?xml version="1.0"?>

<robot name="ai_rover_remastered" xmlns:xacro="http://www.ros.
org/wiki/xacro">

<xacro:macro name="cylinder_inertia" params="m r h">
  <inertial>
    <mass value="${m}"/>

    <inertia ixx="${m*(3*r*r+h*h)/12}" ixy="0" ixz="0"
    iyy="${m*(3*r*r+h*h)/12}" iyz="0" izz="${m*r*r/2}"/>
  </inertial>
</xacro:macro>
```

```
<xacro:macro name="wheel" params="prefix reflect">
  <link name="${prefix}_wheel">
    <visual>
      <origin xyz="0 0 0" rpy="${pi/2} 0 0"/>
      <geometry>
        <cylinder radius="${wheel_radius}" length="0.005"/>
      </geometry>
      <material name="blue"/>
    </visual>

    <collision>
      <origin xyz="0 0 0" rpy="${pi/2} 0 0"/>
      <geometry>
        <cylinder radius="${wheel_radius}" length="0.005"/>
      </geometry>
    </collision>

    <xacro:cylinder_inertia m="10" r="${wheel_radius}"
    h="0.005"/>
  </link>

  <joint name="${prefix}_wheel_joint" type="continuous">
    <axis xyz="0 1 0" rpy="0 0 0" />
    <parent link="base_link"/>
    <child link="${prefix}_wheel"/>
    <origin xyz="${wheel_joint_offset} ${((base_width/2)+base_
    wheel_gap)*reflect} -0.005" rpy="0 0 0"/>
  </joint>
</xacro:macro>

</robot>
```

There are many examples of the object-oriented programming (OOP) paradigm in this short script. The ${prefix} macro creates separate left and right wheel links and joints connected to the base link. You should note the OOP instantiation (left and right wheel joints and links using the ${prefix} macro) and aggregation (connecting the left and right wheels on either side of the base_link). The wheel_joint_offset determines how far horizontally from the base center is the wheel offset.

casterInertia.xacro

Using macros, we can model the mass and moment of inertia for other components on the rover. For instance, we model the caster (casterInertia.xacro) as a "spherical" wheel that rotates in any direction, similar to the physical rover.

```xml
<?xml version="1.0"?>

<robot name="ai_rover_remastered" xmlns:xacro="http://www.ros.
org/wiki/xacro">

<xacro:macro name="sphere_inertia" params="m r">
  <inertial>
    <mass value="${m}"/>
    <inertia ixx="${2.0*m*(r*r)/5.0}" ixy="0.0" ixz="0.0"
    iyy="${2.0*m*(r*r)/5.0}" iyz="0.0" izz="${2.0*m*(r*r)/5.0}"/>
  </inertial>
</xacro:macro>

<link name="caster_wheel">
  <visual>
    <origin xyz="0 0 0" rpy="0 0 0"/>
    <geometry>
      <sphere radius="${caster_wheel_radius}"/>
    </geometry>
```

```
    </visual>

    <collision>
      <origin xyz="0 0 0" rpy="0 0 0"/>
      <geometry>
        <sphere radius="${caster_wheel_radius}"/>
      </geometry>
    </collision>

    <xacro:sphere_inertia m="5" r="${caster_wheel_radius}"/>
</link>

<joint name="caster_wheel_joint" type="continuous">
  <axis xyz="0 1 0" rpy="0 0 0" />
  <parent link="base_link"/>
  <child link="caster_wheel"/>

  <origin xyz="${caster_wheel_joint_offset} 0 -${caster_wheel_
  radius+0.005}" rpy="0 0 0"/>
</joint>

</robot>
```

We can now create a caster wheel with any geometry required to model the physical caster wheel accurately. This source code creates a caster wheel link (caster_wheel) and a joint (caster_wheel_joint) connected to the base link.

laserDimensions.xacro

The Gazebo Laser Range Finding Scanner Plug-in (LRFP) determines the shape and geometry of previously unexplored areas. The LRFP simulates a LiDAR sensor; in our case, the Hokuyo LiDAR system. A LiDAR sensor uses laser pulses to measure distances to objects within the environment, helping to determine their geometry. We can then generate a map, helping the rover to navigate and avoid obstacles. LiDAR systems are one of the primary sources of odometry in modern robotics. The following code (laserDimensions.xacro) places the sensor_laser with its correct geometric dimensions on top of the rover chassis to receive messages from the Xacro sensor file:

```
<?xml version="1.0"?>

<robot name="ai_rover_remastered" xmlns:xacro="http://www.ros.
org/wiki/xacro">
  <xacro:property name="laser_size_x" value="0.03"/>
  <xacro:property name="laser_size_y" value="0.03"/>
  <xacro:property name="laser_size_z" value="0.04"/>
  <xacro:property name="laser_origin_x" value="0.065"/>
  <xacro:property name="laser_origin_y" value="0"/>
  <xacro:property name="laser_origin_z" value="0.035"/>
  <link name="sensor_laser">
    <visual>
      <geometry>
        <box size="${laser_size_x} ${laser_size_y} ${laser_
        size_z}"/>
      </geometry>
      <material name="blue"/>
    </visual>
    <collision>
      <geometry>
        <box size="${laser_size_x} ${laser_size_y} ${laser_
        size_z}"/>
      </geometry>
    </collision>
    <xacro:box_inertia m="0.2" w="${laser_size_x}" h="${laser_
    size_y}" d="${laser_size_z}"/>
  </link>
```

```
<joint name="sensor_laser_joint" type="fixed">
    <origin xyz="${laser_origin_x} ${laser_origin_y} ${laser_
    origin_z}" rpy="0 0 0" />
    <parent link="base_link"/>
    <child link="sensor_laser" />
</joint>
</robot>
```

cameraDimensions.xacro

The integrated rover camera captures and processes image data to sense and avoid obstacles. For simplicity, we make the camera dimensions the same as the LiDAR sensor and place it in front of and below the LiDAR sensor. The cameraDimensions.xacro defines a simple camera housing with a fixed camera_link in front of the chassis (base_link):

```
<?xml version="1.0"?>

<robot name="ai_rover_remastered" xmlns:xacro="http://www.ros.
org/wiki/xacro">
    <xacro:property name="camera_size_x" value="0.03"/>
    <xacro:property name="camera_size_y" value="0.03"/>
    <xacro:property name="camera_size_z" value="0.04"/>
    <xacro:property name="camera_origin_x" value="0.165"/>
    <xacro:property name="camera_origin_y" value="0"/>
    <xacro:property name="camera_origin_z" value="-0.035"/>
    <link name="camera_link">
        <visual>
            <geometry>
                <box size="${camera_size_x} ${camera_size_y}
                ${camera_size_z}"/>
            </geometry>
            <material name="blue"/>
```

```
        </visual>
        <collision>
            <geometry>
                <box size="${camera_size_x} ${camera_size_y}
                ${camera_size_z}"/>
            </geometry>
        </collision>
        <xacro:box_inertia m="0.2" w="${camera_size_x}"
        h="${camera_size_y}" d="${camera_size_z}"/>
    </link>
    <joint name="camera_joint" type="fixed">
        <origin xyz="${camera_origin_x} ${camera_origin_y}
        ${camera_origin_z}" rpy="0 0 0" />
        <parent link="base_link"/>
            <child link= "camera_link" />
    </joint>
</robot>
```

IMUDimensions.xacro

In this section, we define the inertial measurement unit (IMU) plug-in. The IMU data captures the rover's speed (straight and turning) and orientation (attitude) relative to the environment. The IMU heading and attitude data is processed so the rover can maneuver in an environment. The SLAM process needs the IMU and wheel encoder data to help accurately and precisely outline boundaries, walls, and obstacles (Chapter 7).

The IMU housing dimensions will be small compared to those of the camera and LiDAR. To simplify the physics, the IMU location will be in the origin (xyz=0,0,0) of the rover. To avoid the LiDAR sweep, we reduce the

IMU's size. The `IMUDimensions.xacro` defines a simple IMU housing with a fixed `IMU_link` relative to the chassis (`base_link`):

```xml
<?xml version="1.0"?>
<robot name="ai_rover_remastered" xmlns:xacro="http://www.ros.
org/wiki/xacro">
  <xacro:property name="IMUsizeX" value="0.01"/>
  <xacro:property name="IMUsizeY" value="0.01"/>
  <xacro:property name="IMUsizeZ" value="0.01"/>
  <xacro:property name="IMUoriginX" value="0"/>
  <xacro:property name="IMUoriginY" value="0"/>
  <xacro:property name="IMUoriginZ" value="0.015"/>
  <link name="IMU_link">
    <visual>
      <geometry>
        <box size="${IMUsizeX} ${IMUsizeY} ${IMUsizeZ}"/>
      </geometry>
      <material name="blue"/>
    </visual>
    <collision>
      <geometry>
        <box size="${IMUsizeX} ${IMUsizeY} ${IMUsizeZ}"/>
      </geometry>
    </collision>
    <xacro:box_inertia m="0.2" w="${IMUsizeX}" h="${IMUsizeY}"
    d="${IMUsizeZ}"/>
  </link>
  <joint name="IMU_joint" type="fixed">
    <origin xyz="${IMUoriginX} ${IMUoriginY} ${IMUoriginZ}"
    rpy="0 0 0" />
```

```
    <parent link="base_link"/>
    <child link= "IMU_link" />
  </joint>
</robot>
```

Gazebo Plug-ins

A plug-in is a section of source code compiled as a C++ library and inserted into any Gazebo simulation. Python is an "interpreted" language (slow), while C++ is compiled (fast). All plug-ins have direct access to all the functions of the physics engine of Gazebo.

Plug-ins are helpful because they

- let developers control and enhance Gazebo features;

- are self-contained software routines for simulations; and

- can be inserted and removed from a running system.

Previous versions of Gazebo utilized integrated controllers, which behaved in much the same way as modern Gazebo plug-ins. Consequently, no enhancements were possible with these controllers. Current Gazebo plug-ins are now far more flexible and allow users to program what functionality to include in their simulations.

You should only use a plugin when

- you want to alter a simulation programmatically, such as responding to simulation events; and

- you want a fast interface to Gazebo without the overhead of the transport layer, such as an interface. An example of an interface would be to control the speed and direction of our rover.

Plug-in Types

There are currently six types of Gazebo plug-ins, as follows:

1. World (catacombs, etc.)

2. Robot Model (rover)

3. Sensor (LiDAR, IMU, camera, etc.)

4. System (differential-drive controller, etc.)

5. Visual (laser view [blue field in Figure 5-8], etc.)

6. GUI (rover controls)

Each plug-in is attached to a specific "object" in the Gazebo environment. For example, the robot model plug-in is attached to and controls the rover in Gazebo. Similarly, the world plug-in is attached to a catacombs environment, and each sensor has a sensor plug-in. The system plug-in is specified in the command line and loads the wheels and caster physics configuration in the differential-drive controller. The visual plug-in is automatically loaded and shows colors as defined in the different Xacro files; e.g., the "blue" wheels, "red" LiDAR box, etc. The GUI plug-in is also automatically loaded and connects the Gazebo controls to control objects (and sub-objects such as joints) in the environment (move, turn, rotate, etc.). These controls ARE NOT the same as simulation controls using Teleops.

Differential-Drive Controller (DDC) Plug-in

The differential-drive controller (DDC) plug-in is a system plug-in that ties the physics engine to the rover. The DDC uses the `wheels.xacro` definitions to attach the individually defined parts of the DDC to the physics engine. This connection will be used by the Teleops keystrokes to realistically control the movement of the rover.

```
<gazebo>
  <plugin name="differential_drive_controller"
filename="libgazebo_ros_diff_drive.so">
    <alwaysOn>false</alwaysOn>
    <legacyMode>false</legacyMode>
    <updateRate>20</updateRate>
    <leftJoint>left_wheel_joint</leftJoint>
    <rightJoint>right_wheel_joint</rightJoint>
    <wheelSeparation> ${wheel_separation}
    </wheelSeparation>
    <wheelDiameter>${wheel_radius * 2}
    </wheelDiameter>
    <torque>20</torque>
    <commandTopic>
      /ai_rover_remastered/base_controller/cmd_vel
    </commandTopic>
    <odometryTopic>
      /ai_rover_remastered/base_controller/odom
    </odometryTopic>
    <odometryFrame>odom</odometryFrame>
    <robotBaseFrame>base_footprint</robotBaseFrame>
  </plugin>
</gazebo>
```

In this section, we integrate the DDC plug-in into the remastered rover model. We do this to design, develop, and test the manual controls of the rover. The simple manual controls move the rover forward or backward and turn it left or right. Because the wheels turn independently, their relative velocities can be different, too. These manual controls test the differential drive so that when the autonomous navigation control is added, the control logic is correct. Furthermore, this initial testing of the differential drive reinforces the basic ROS control concepts.

We do this by integrating the built-in DDC plug-in into the Xacro rover model. Create a script file called ai_rover_remastered_plugins.xacro in the current directory with the following content:

```
<?xml version="1.0" ?>
<robot name="ai_rover_remastered" xmlns:xacro="https://www.ros.
org/wiki/xacro" >
  <!--
```

Now we are ready to add control to our robot. We will add a new plug-in to our Xacro file, and we will add a differential-drive plug-in to our robot. The new tag looks as follows:

```
  -->
<gazebo>
<plugin name="differential_drive_
controller"    filename="libgazebo_ros_diff_drive.so">
<alwaysOn>false</alwaysOn>
<legacyMode>false</legacyMode>
<updateRate>20</updateRate>
<leftJoint>left_wheel_joint</leftJoint>
<rightJoint>right_wheel_joint</rightJoint>
<wheelSeparation>${wheel_separation}</wheelSeparation>
<wheelDiameter>${wheel_radius * 2}</wheelDiameter>
<torque>20</torque>
<commandTopic>/ai_rover_remastered/base_controller/cmd_vel</
commandTopic>
<odometryTopic>/ai_rover_remastered/base_controller/odom</
odometryTopic>
<odometryFrame>odom</odometryFrame>
<robotBaseFrame>base_footprint</robotBaseFrame>
</plugin>
</gazebo>
```

Let us briefly review this file. The first line we will examine is the `filename` line. The filename is the `gazebo_ros` library name containing the plug-in implementation for the differential-drive controller. We will quickly review the following defined tags for this plug-in:

- The plug-in name is `differential_drive_controller` and is located in the library `libgazebo_ros_diff_drive.so`.

- The `<alwaysOn>` tag allows the robot to receive velocity commands; default is set to `false`.

- The `<legacyMode>` tag is `false`, not allowing us to swap the left and right wheels.

- The `<updateRate>` tag is 20 Hz, the frequency of information sent to the controller.

- The `<leftJoint>` tag is the name of the left joint.

- The `<rightJoint>` tag is the name of the right joint.

- The `<wheelSeparation>` tag is the distance from the center of one wheel to the center of the other in meters. Usually this defaults to 0.34 meters.

- The `<wheelDiameter>` tag is the diameter of each wheel. Usually, the diameter of each wheel is the same and is set as a default to 0.15 meters.

- The `<commandTopic>` tag is to receive `geometry_msgs` or `Twist` message commands from the user or deep learning or cognitive AI control architectures.

- The `<odometryTopic>` tag is to publish nav_msgs or odometry messages.

- The <odometryFrame> tag defaults to the odometry frame.

- The <robotBaseFrame> tag is the rover frame to calculate odometry from and will default to base_footprint.

The <plugin> block and definitions must be in a <gazebo> block when creating a plug-in script.

Finally, the plug-in publishes the odometry, <odom>, of the rover. Odometry uses sensor data (wheel encoders) to estimate the rover's change in position over time and calculates the current position of the rover relative to its starting location. Each movement of the rover triggers a sensor reading that is used to update the internal map location. This calculation can be extremely sensitive to errors and sources of uncertainties caused by the integration of velocity measurements over time.

Laser Plug-in

The laser plug-in is a sensor plug-in. It is a generic laser modified to the specifications of our actual laser defined in laserDimensions.xacro.

Now that we have added the DDC plug-in and the geometry of the simulated LiDAR sensor component to the chassis, we have to connect it to the LRFP. The LFRP provides the internal logic, behavior, and characteristics of the Hokuyo LiDAR system. The following code accesses the Gazebo plug-in library for the LRFP and sets the LiDAR sensor to Hokuyo characteristics:

```
<!--Gazebo Hokuyo Laser Plugin-->
<gazebo reference="sensor_laser">
    <sensor type="ray" name="head_hokuyo_sensor">
        <pose>0 0 0 0 0 0</pose>
        <visualize>true</visualize>
```

```
        <update_rate>20</update_rate>
        <ray>
            <scan>
                <horizontal>
                    <samples>1440</samples>
                    <resolution>1</resolution>
                    <min_angle>-3.14159</min_angle>
                    <max_angle>3.14159</max_angle>
                </horizontal>
            </scan>
            <range>
                <min>0.10</min>
                <max>30.0</max>
                <resolution>0.01</resolution>
            </range>
            <noise>
                <type>gaussian</type>
                <mean>0.0</mean>
                <stddev>0.01</stddev>
            </noise>
        </ray>
        <plugin name="gazebo_ros_head_hokuyo_controller"
        filename="libgazebo_ros_laser.so">
            <topicName>/ai_rover_remastered/laser_scan/scan
            </topicName>
            <frameName>sensor_laser</frameName>
        </plugin>
        </sensor>
</gazebo>
```

The LiDAR sensor system publishes the laser scan data to the /ai_
rover_remastered/sensor_laser/scan topic. The TF frame subscribes to

the sensor_laser link, which integrates the LiDAR sensor model with the rest of the rover model. (If we also change our LiDAR sensor system from Hokuyo, we must alter the range, sample_rate, min_angle, max_angle, resolution, and signal_noise parameters to match the new LiDAR system.) After integrating the link-and-joint Xacro URDF and the LiDAR plug-in code, run the updated rover model with the following terminal shell command:

**$ roslaunch ai_rover_remastered ai_rover_remastered_
gazebo.launch**

After running, you should have a display similar to Figure 5-2. Notice that the sensor has an infinite range and can see 360° in a horizontal circle of one pixel thick, but it has a blind spot close to the sensor. The LiDAR cannot detect any objects in this blind spot.

Figure 5-2. *Rover and LiDAR sensor: Blue area is the sensor coverage. Gray circle is the LiDAR blind spot*

Placing a cube object in the environment shows a gray area behind the cube (Figure 5-3). This gray area is another blind spot. The blue areas are the LiDAR sensor sweep showing no object encountered. Again, the sensor has an infinite range.

Figure 5-3. *Rover and LiDAR sensor operating (Gazebo)*

The LiDAR sensor publishes the data in the RViz environment under the sensor_laser/scan topic (Figure 5-4).

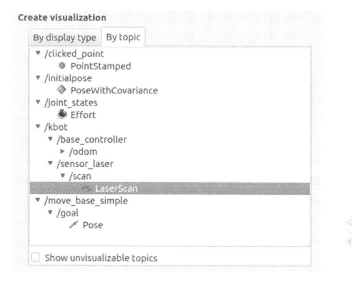

Figure 5-4. *LiDAR senor scan topic*

From the rover's point of view, the LiDAR sensor "sees" the boundary of the cube shape as a thick red line (Figure 5-5). The LiDAR sensor publishes that the laser has intersected an object.

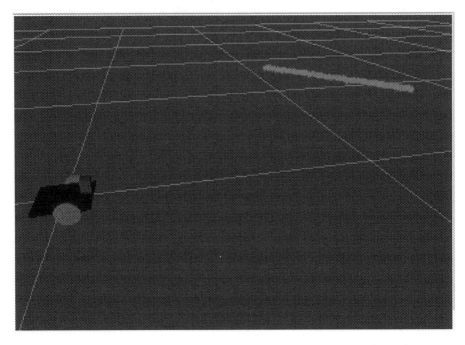

Figure 5-5. *Rover "seeing" the cube as a boundary line (RViz)*

Now we will integrate the Gazebo camera plug-in for the rover.
This plug-in will allow us to view what the rover "sees" as it explores
the environment. Our simulated camera in ROS provides image data
to the rover for object identification, tracking, and manipulation tasks.
Noetic ROS currently supports both monocular and stereo cameras. To
simplify this book, we will only use a simulated monocular camera for
our ROS/Gazebo/RViz setup and only a physical monocular camera
for the GoPiGo3 rover. We could use stereo cameras to create SLAM
(Simultaneous Localization and Mapping) of our environment, but this
would only add complexity to this book. We only need the monocular
camera to locate and identify objects that the front of the rover is facing in
an environment. We will only be using the LiDAR/SLAM configurations to
sense and avoid obstacles and locate anomalies in an environment for the
rover to explore further and examine.

Camera Plug-in

The camera plug-in is a sensor plug-in that connects the simulated camera images from Gazebo, then displays them in RViz. The RViz GUI can now "display" what the rover now "sees" in the Gazebo environment.

```
<!-- camera -->
  <gazebo reference="camera_link">
    <sensor type="camera" name="camera1">
      <update_rate>30.0</update_rate>
      <camera name="head">
        <!--
        <horizontal_fov>1.3962634</
         horizontal_fov> -->
        <horizontal_fov>1.57</horizontal_fov>
        <!- ~90 degrees -->
        <image>
          <width>800</width>
          <height>800</height>
          <format>R8G8B8</format>
        </image>
        <clip>
          <near>0.02</near>
          <far>300</far>
        </clip>
        <noise>
          <type>gaussian</type>
          <!-- Noise is sampled independently per
          pixel on each frame. Adding pixel's noise
          value to each color
          channels, which at that point lie in the
          range [0,1]. -->
```

```
        <mean>0.0</mean>
        <stddev>0.007</stddev>
      </noise>
    </camera>
    <plugin name="camera_controller"
     filename="libgazebo_ros_camera.so">
        <alwaysOn>true</alwaysOn>
      <updateRate>0.0</updateRate>
      <cameraName>ai_rover_remastered/camera1</
        cameraName>
      <imageTopicName>image_raw</imageTopicName>
      <cameraInfoTopicName>camera_info</
        cameraInfoTopicName>
      <frameName>camera_link</frameName>
      <hackBaseline>0.07</hackBaseline>
      <distortionK1>0.0</distortionK1>
      <distortionK2>0.0</distortionK2>
      <distortionK3>0.0</distortionK3>
      <distortionT1>0.0</distortionT1>
      <distortionT2>0.0</distortionT2>
    </plugin>
  </sensor>
</gazebo>
```

Once we have inserted the Gazebo camera plug-in in the ai_rover_ remastered_plugins.xacro, we should make certain that the camera plugin is indeed operational. This requires that we set the correct parameters in RViz to accept the incoming image_raw from the Gazebo simulation, which requires that we first start the Gazebo simulation launch file, followed by the RViz launch file. Please look to the Gazebo and RViz launch file sections of this chapter. To accept the image_raw from the Gazebo simulation, we must first go to the display options and select Add, and then select the camera option. Next, add /ai_rover_remastered/ camera1/image_raw to the image topic menu for the camera under the Displays menu. The rover now "sees" a spherical object in Gazebo by "seeing" that same spherical object from the perspective of the rover in the RViz Camera window (Figure 5-6).

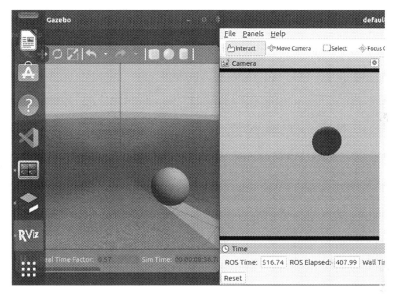

Figure 5-6. *Rover's camera "seeing" the sphere in Gazebo (left) and Rviz (right)*

IMU Plug-in

The IMU plug-in is a sensor plug-in. It connects the rover location to the global environment. Think of the rover as having a local coordinate system, while the environment has a global coordinate system. When the rover moves "1 space forward" in its local system, it moves from <x,y> to <x+1, y> in the global environment. The plug-in also maps the internal acceleration to the global environment to see it in Gazebo (purple arrow).

An IMU must aide the robot's navigation tasks to allow genuine autonomy for the rover. The IMU (inertial measurement unit) sensor must measure and report the rover's speed (accelerometer), direction,

193

acceleration, specific force, angular rotation rate (gyroscope), and the magnetic field (magnetometer) surrounding the rover in all three directions (x, y, and z). We will need both the IMU and wheel encoder values to estimate the robot's 6D pose and position in maps generated with Simultaneous Localization and Mapping (SLAM). The IMU can also combine input from several different sensor types to estimate output movement accurately.

We do this by integrating the IMU plug-in into the Xacro rover model. Create a script file called `ai_rover_remastered_plugins.xacro` in the current directory with the following content:

```
<gazebo>
    <plugin name="imu_plugin"
      filename="libgazebo_ros_imu.so">
      <alwaysOn>true</alwaysOn>
      <bodyName>IMU_link</bodyName>
      <topicName>imu</topicName>
      <serviceName>imu_service</serviceName>
      <gaussianNoise>0.0</gaussianNoise>
      <updateRate>20.0</updateRate>
    </plugin>
</gazebo>
```

Let us briefly review this file. The first line we will examine is the `filename` line. The filename is the `libgazebo_ros_imu.so` library and contains the plug-in implementation for the IMU sensor plug-in. We will quickly review the following defined tags for this plug-in:

- The plug-in name is imu_plugin and is located in the library `libgazebo_ros_diff_drive.so`.

- The `<alwaysOn>` tag allows the IMU to send data.

- The `<bodyName>` tag is set to `IMU_link`, the IMU object link, a child link to `base_link`, which is the rover chassis.

- The `<topicName>` tag is the message tag from IMU.

- The ‹serviceName› tag is the message
  from the IMU_service

- The ‹gaussianNoise› tag is set to the value of zero,
  which means there is no gaussian noise to the LIDAR
  sensor simulations. This might need to be changed to
  closely reflect the explored environment.

- The ‹updateRate› tag is the frequency of sensor
  updates for the LIDAR, which in the case of the source
  code is set to 20 hertz.

The ‹plugin› block and definitions must be in a ‹gazebo› block when
creating a plug-in script.

Visuals Plug-in

The visuals plug-in is (obviously) a visual plug-in. The material colors
defined in the Xacro files are only used in RViz. The colors have to be
redefined in our visuals.xacro plug-in to be displayed Gazebo.

```
<?xml version="1.0" ?>
<robot name="ai_rover_remastered" xmlns:xacro="https://www.ros.
org/wiki/xacro" >
  <!-- Define color for robot parts -->
  <gazebo reference="base_link">
    <material>Gazebo/Orange</material>
  </gazebo>
  <gazebo reference="left_wheel">
    <material>Gazebo/Blue</material>
  </gazebo>
  <gazebo reference="right_wheel">
    <material>Gazebo/Blue</material>
  </gazebo>
</robot>
```

Putting It All Together

We have now defined the individual components of the physical rover and their associated plug-ins for Gazebo. To complete the construction of our rover, we need to "glue" them together. First, we create the plug-in file (ai_ rover_remastered_plugins.xacro) and then the rover model (ai_rover_ remastered.xacro), which includes the plug-ins.

ai_rover_remastered_plugins.xacro

Now that we have created our plug-ins for each rover component, we have to combine them in a single module and load them into Gazebo and RViz. To do this, we "include" the individual plug-in files into an ai_rover_ remastered_plugins.xacro file.

```
<?xml version="1.0" ?>
<robot name="ai_rover_remastered" xmlns:xacro="https://www.ros.
org/wiki/xacro" >
  <xacro:include filename="$(find ai_rover_remastered)/urdf/
  visuals.xacro"/>
  <xacro:include filename="$(find ai_rover_remastered)/urdf/
  DDC_plugin.xacro"/>
  <xacro:include filename="$(find ai_rover_remastered)/urdf/
  Laser_plugin.xacro"/>
  <xacro:include filename="$(find ai_rover_remastered)/urdf/
  Camera_plugin.xacro"/>
  <xacro:include filename="$(find ai_rover_remastered)/urdf/
  IMU_plugin.xacro"/>
</robot>
```

ai_rover_remastered.xacro

Finally, we bring all of these separate components together in the
ai_rover_remastered.xacro file. This file "glues" or constructs the
individual parts (chassis, wheels, caster, etc.) into a cohesive whole and
complete rover.

```xml
<?xml version="1.0"?>

<robot name="ai_rover_remastered" xmlns:xacro="http://www.ros.
org/wiki/xacro">

  <xacro:include filename="$(find ai_rover_remastered)/urdf/
  dimensions.xacro"/>
  <xacro:include filename="$(find ai_rover_remastered)/urdf/
  chassisInertia.xacro"/>
  <xacro:include filename="$(find ai_rover_remastered)/urdf/
  wheelInertia.xacro"/>
<xacro:include filename="$(find ai_rover_remastered)/urdf/
casterInertia.xacro"/>
<xacro:include filename="$(find ai_rover_remastered)/urdf/
laserDimensions.xacro"/>
<xacro:include filename="$(find ai_rover_remastered)/urdf/
cameraDimensions.xacro"/>
<xacro:include filename="$(find ai_rover_remastered)/urdf/ ai_
rover_remastered_plugins.xacro"/>
</robot>
```

Now that we have remastered the rover into modules using Xacro, we
need to convert it to an URDF file and then verify the converted file has
no errors. To do this, we run the following two terminal commands in the
URDF directory:

```
$ rosrun xacro xacro ai_rover_remastered.xacro > rover.urdf
$ check_urdf  rover.urdf
```

Assuming everything is correct, we should have the following check_ urdf output:

```
Robot name is: ai_rover_remastered
-----------Successfully Parsed XML ------------------
Root Link: base_footprint has 1 child(ren)
    child(1): base_link
            child(1): caster_wheel
            child(2): left_wheel
            child(3): right_wheel
```

If there are any errors, go back and review the scripts for typos.

RViz Launch File

Recall that RViz is used to simulate the rover in isolation; i.e., we see the world "through the eyes" of the rover. To prepare the project for RViz, we convert the ai_rover_remastered directory into an ROS package just like we did in Chapter 4.

```
$ cd ~/catkin_ws/src/
$ catkin_create_pkg ai_rover_remastered
```

The catkin_create_pkg command creates two files in the ai_rover_ remastered directory: CMake_lists.txt and package.xml. Finally, create the ai_rover_remastered_rviz.launch file in the launch directory:

```
<launch>
<param name="robot_description" command="$(find xacro)/xacro
-- inorder
$(find ai_rover_remastered)/urdf/ai_rover_remastered.xacro" />
<node name="robot_state_publisher" pkg="robot_state_publisher"
type="robot_state_publisher"/>
```

```
<node name="joint_state_publisher" pkg="joint_state_publisher"
type="joint_state_publisher"/>
<node name="rviz" pkg="rviz" type="rviz" required="true"/>
</launch>
```

This launch file is nearly identical to the RViz launch file in Chapter 4, the only difference being the script files with the Xacro extension. Similar to Chapter 4, Noetic ROS launches three nodes: robot_state_publisher, joint_state_publisher, and rviz. The first two guarantee the correct transformation between the link and joint frames published by ROS. The final ROS node launches the RViz program. Just as we did in Chapter 4, compile the code using the following:

```
$ cd catkin_ws/
$ catkin_make
$ source devel/setup.sh
$ roslaunch ai_rover_remastered ai_rover_remastered_RViz.launch
```

The roslaunch command launches the RViz environment and GUI. To eliminate the errors found within RViz, we will need to alter the *Global Options* ➤ fixed_frame ➤ *map* to *Global Options* ➤ fixed_frame ➤ *base_link*. Therefore, just like in Chapter 4, we will have to add the RobotModel within the Displays tab (Figure 5-7).

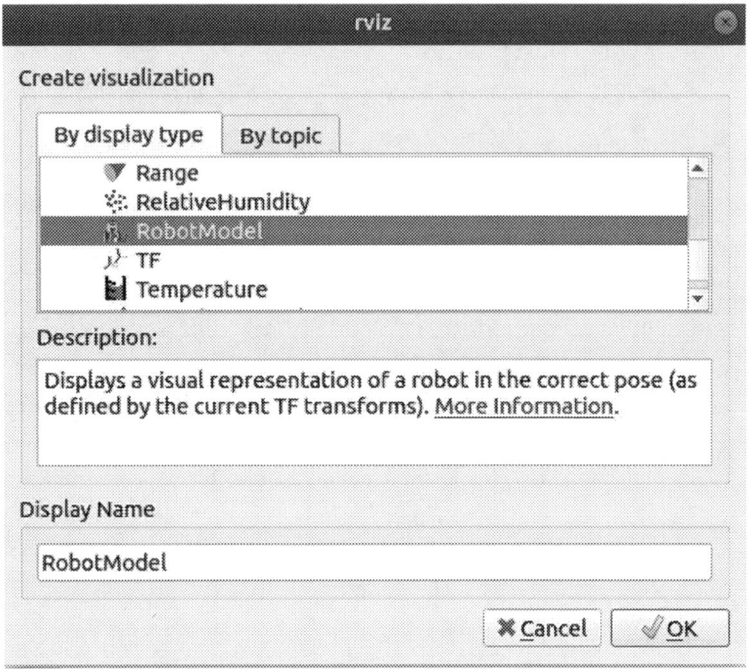

Figure 5-7. Adding the Xacro RobotModel within Rviz

Gazebo Launch File

Recall that Gazebo is used to test the rover in a robust environment. That means we see the rover and background from a third-person perspective. The rover, once verified to be correct in RViz, needs to be imported to Gazebo, and Gazebo needs two changes:

1. Convert the Xacro files to URDF files. Gazebo "gets confused" reading valid Xacro files, but converting them to a single URDF removes the most warnings.

    ```
    $ rosrun xacro xacro ai_rover_remastered.xacro >
    ai_rover_remastered.urdf
    ```

2. Copy the ai_rover_gazebo.launch file from Chapter 4 and call it ai_rover_remastered_gazebo. launch. Change any occurrence of <ai_rover> to <ai_rover_remastered>.

```
<launch>
  <!-- We resume the logic in
empty_world.launch, changing only the name of the world
to be launched -->

<include file="$(find gazebo_ros)/launch/empty_world.
launch">
<arg name="world_name" value="$(find ai_rover_
remastered)/worlds/MYROBOT.world"/>

<!-- more default parameters can be changed here -->
</include>
<node name="spawn_urdf" pkg="gazebo_ros" type="spawn_
model" args="-file $(find ai_rover_remastered)/urdf/
ai_rover_remastered.urdf -urdf -x 0 -y 0 -z 0 -model
ai_rover_remastered" />
</launch>
```

The MYROBOT.world file is the same one we used in Chapter 4.

The rover model spawned in Gazebo uses the spawn_model node of the gazebo_ros package. The rover model passes as an argument to the Gazebo instance. The following is the roslaunch terminal command that launches the rover model in Gazebo:

```
$ roslaunch ai_rover_remastered ai_rover_remastered_
gazebo.launch
```

> **Note**   Before launching a Gazebo simulation, go to the home directory and kill all Gazebo and Gzserver processes:
>
> ```
> $ killall gazebo
> ```
> ```
> $ killall gzserver
> ```
>
> These commands guarantee a clean slate for Gazebo simulations. If you get the error `[Err] [REST.cc:205] Error in REST request`, refer to `https://automaticaddison.com/how-to-launch-gazebo-in-ubuntu/`.

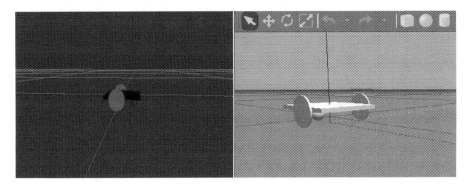

Figure 5-8. *Remastered rover displayed in Rviz and Gazebo*

Once we execute the RViz and Gazebo launch commands, we will have the following RViz and Gazebo displays (Figure 5-8).

Troubleshooting Xacro and Gazebo

This section reviews issues you might encounter while developing the Xacro model and converting it to URDF for Gazebo.

- Gazebo has two visual components: the environment and the rover. The potential "physical" interaction of the rover and the environment could cause your program (or operating system) to crash. Verify that the spawning of the rover does not overlap any objects in the environment.

- Run the simulation once, save the configuration using the *Global ➤ Options* menu, and then exit. Now the saved configuration will automatically be loaded every time you run Gazebo.

- Make sure to terminate ALL Gazebo and Gzserver processes. A fresh start of these processes will ensure fewer problems.

Teleop Node for Rover Control

Teleop means to control (**op**erate) from a distance (tele-). One of the ROS packages that functions to remotely control the rover is teleop_twist_ keyboard. This function uses the keys i/j/l/ to move the rover up/left/ right/down, respectively. The teleop_twist_keyboard intercepts the keyboard commands and passes (publishes) the information via the cmd_ vel topic to all subscribers. We bind the DDC plug-in to the cmd_vel topic by subscribing.

We will use two terminals to install the Teleops package. In the first terminal:

```
$ sudo apt-get install ros-noetic-teleop-twist-keyboard
$ roscore
```

In the second terminal:

```
$rosrun teleop_twist_keyboard teleop_twist_keyboard.py
```

Optionally, in a third terminal, you can view the published messages for all subscribers:

```
rostopic echo /cmd_vel
```

As usual, we prepare the groundwork for our Teleops control package by creating the Ubuntu directory structure and placing ourselves in the correct directory:

```
$ cd catkin_ws/src
$ catkin_create_pkg ai_rover_simple_control
$ cd ai_rover_simple_control
$ mkdir -p launch src
$ cd ~/catkin_ws
$ catkin_make
$ source devel/setup.sh
$ cd catkin_ws/src/ai_rover_simple_control/launch
```

Next, create the ai_rover_teleops.launch file that publishes the cmd_vel topic:

```
<?xml version="1.0"?>
<launch>
<node name="teleop" pkg="teleop_twist_keyboard" type="teleop_
twist_keyboard.py" output="screen">
<remap from="/cmd_vel" to="/ai_rover_remastered/base_
controller/cmd_vel"/>
</node>
</launch>
```

Now, make this launch file executable:

```
$ chmod +rwx ai_rover_teleops.launch
```

Transform (TF) Graph Visualization

In the context of this project, a transform is a single-movement step in time. Each movement generates messages to each component to reflect the location change. Messages passed between nodes are difficult to visualize without a tool. We use the TF Graph tool to visualize (and test) the connections between the teleops_twist_keyboard package and the DDC plug-in. Using TF Graph, we see the real-time published messages, and we can use this information to guide our debugging.

First, we must get our rover running in Gazebo (Figure 5-9). Create three side-by-side terminals in the Terminator program. Launch the Gazebo program (not shown) in the left terminal (orange box):

```
$ roslaunch ai_rover_remastered ai_rover_remastered_
gazebo.launch
```

In the middle terminal, launch the teleops launch file (blue box):

```
$ roslaunch ai_rover_remastered_simple_control ai_rover_
remastered_teleops.launch
```

Be careful using the keyboard; any key press could now accidentally be interpreted as a teleop command! Click your mouse in the right terminal (green box) to make it active. In the right terminal run the following:

```
$ cd ~/catkin_ws/
$ ~/catkin_ws/ source devel/setup.sh
$ ~/catkin_ws/ rosrun tf2_tools view_frames.py
```

After completing the previous set of commands, your screen should look like Figure 5-9.

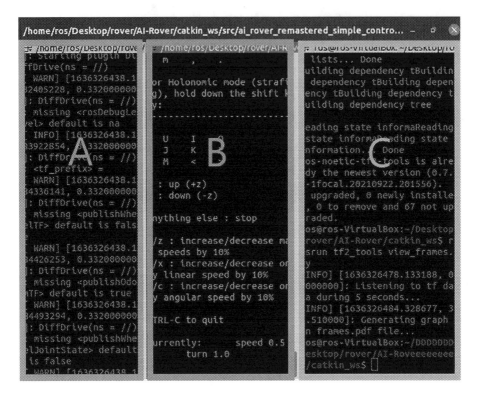

Figure 5-9. *Multiple terminals are running the tf2_tools ROS analysis program*

Once we have determined that the ROS program is working, we can control the rover directly in Gazebo (Figure 5-10).

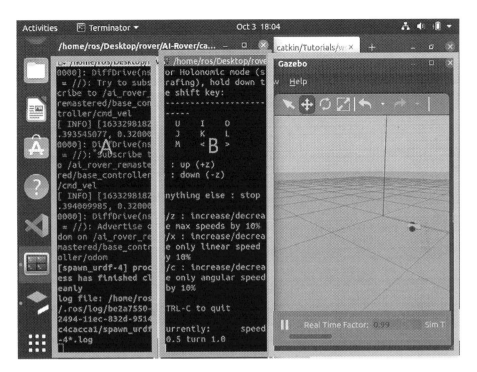

Figure 5-10. *Multiple terminal windows and Gazebo running*

Please note that once these terminals are running, we need to be certain that the terminal responsible for executing the teleops.py program (blue box) is active by clicking on it with our pointer. To review: The launch files are the orange box, the active terminal found in blue box performs the teleops.py program, and the TF View_Frames is the green box.

The rosrun tf view_frames command produces a pop-up window with the TF Graph (Figure 5-11) showing the interrelation between the rover components. Each component has four properties: broadcaster, average rate, recent transform, and buffer length. The broadcaster is the package that published the data, and the average rate is the frequency of updates. The recent transform is the internal clock timestamp for the last update. And, finally, the buffer length is the amount of time that transpired to complete the previous update. Of these four properties, we will only be using broadcaster.

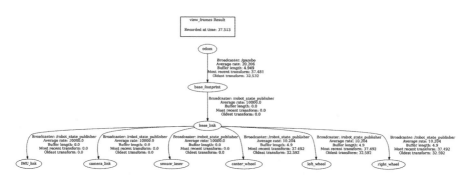

Figure 5-11. *Frames outline for both rover URDF and Gazebo*

Troubleshooting RViz Window Errors

In Figure 5-12, if you are missing the RobotModel field or TF field, you need
to add it. Luckily, both solutions are very similar. Go to the Add button
and select RobotModel or TF. Expand the TF visualization to see coordinate
frames; these are our components defined in our Xacro files. Select the
checkbox for the RobotModel and TF displays. These should now be shown
in Gazebo and be reflected in the RViz GUI panel.

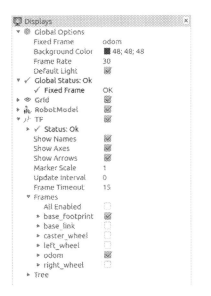

Figure 5-12. *The correct settings for the RViz Displays panel*

To reflect these changes in the RViz Views panel, change orbit view in the Views panel to odom: Current View ➤ Target Frame ➤ odom. See Figure 5-13.

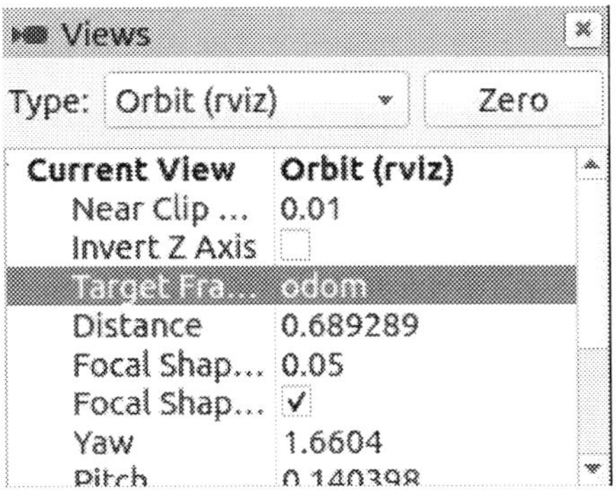

Figure 5-13. *Odometry setting for the Target Frame option*

Controlling the Rover

We can control the rover in Gazebo with keyboard commands. Click on the Terminator terminal to interact with the `teleops_twist_keyboard` program (Figure 5-9, blue box). Press the "i" key, and the rover will continuously move forward in both RViz and Gazebo until you tell it to stop. Some simple case-sensitive keyboard commands: "i" (move forward), "j" (turn left), "k" (stop), "l" (turn right). Other commands may be defined as we need them. Figure 5-14 shows a snapshot of the rover in the RViz and Gazebo environments.

Figure 5-14. *Rover in Rviz (left) and Gazebo (right)*

Remember to save your work: File ➤ "Save Config As." Save the configuration file in the following directory:

`ai_rover_remastered_simple_control/config/`
`rviz_odom.rviz`

Reload the program with the new configuration file.

Drifting Issues with the Rover

As you turned and moved the rover in the world, there may have been "drifting" problems. This drifting means the rover does not travel in a straight line. Usually, drifting occurs after the rover turns. If this happens, alter the mass of the rover by modifying the corresponding inertia values in the Xacro URDF file. Experiment with the weights of the different components until you are satisfied. These changes should control the drifting issues with the rover to a greater extent. Eventually, we will use deep learning to prevent drifting errors. For additional information, refer to `https://www.youtube.com/watch?v=1bnEdQzf8Yw`.

Our First Python Controller

The DDC plug-in subscribes to the `cmd_vel` topic to receive velocity commands to control the rover. The DDC does not know where the data originates. In Chapter 4, keyboard commands (i/j/k/l) published messages to the `cmd_vel` topic. But we can send commands from a program! The `Twist` library function publishes command messages to the `cmd_vel` topic, too. But because it accepts parameters, command control and accuracy are better than the keyboard version.

The `Twist` function has two velocity attributes: linear (forward/backward) and angular (turning). The parameters set these attributes. For instance, in the following Python script, set the `msg.linear.x=0.1` initial value for `Twist()`. This parameter translates to "move the Rover forward 0.1m every second" and is equivalent to pressing the "i" key once on the keyboard.

```
ai_rover_remastered_simple_control/src /ai_rover_simple_
twist_pub.py
```

Create the ai_rover_remastered_simple_twist_pub.py script and make it executable:

```python
#!/usr/bin/env python3
import rospy
import sys
from geometry_msgs.msg import Twist

def publish_velocity_commands():

  # Velocity publisher

  vel_pub = rospy.Publisher('/ai_rover_remastered/base_
  controller/cmd_vel', Twist, queue_size=10)

  rospy.init_node('ai_rover_simple_twist_pub', anonymous=True)

  msg = Twist()
  msg.linear.x = 0.1
  msg.linear.y = 0
  msg.linear.z = 0
  msg.angular.x = 0
  msg.angular.y = 0
  msg.angular.z = 0

  rate = rospy.Rate(10) # 10hz
  while not rospy.is_shutdown():
    vel_pub.publish(msg)
    rate.sleep()

if __name__ == '__main__':
  if len(sys.argv) == 1:
    try:
      publish_velocity_commands()
```

Load the configuration file from the first simple control keyboard program before executing the Python control program. If you do not, then you may encounter difficulties. Using the Terminator program, launch the RViz and Gazebo simulators:

```
$ roslaunch ai_rover_remastered ai_rover_remastered_
gazebo.launch
$ chmod +rwx src/ai_rover_remastered_simple_control/src/ai_
rover_remastered_simple_twist_pub.py
$ rosrun ai_rover_remastered_simple_control ai_rover_
remastered_simple_twist_pub.py
```

The ai_rover_remastered_simple_twist_pub.py controller script publishes a Twist message directly to the cmd_vel topic to which the DDC subscribes. Experiment with the movement for the Twist function by changing any of the msg.linear or msg.angular fields. You should be able to observe the rover moving around in the RViz and Gazebo environments. We now have ALL the fundamental building blocks needed to construct an autonomous rover.

Building Our Environment

After we have created the controller script for the rover, it moves without human input. We have now developed a primitive autonomous rover, and now we have to put it somewhere to explore.

We will simulate our environment as a simple maze. Go to the Gazebo GUI, select Edit (Open Building Editor), and create a simple maze with a few walls and at least one door. For more information, refer to http://gazebosim.org/tutorials?cat=build_world&tut=building_editor. Our generated maze is shown in Figure 5-15; yours can be different. We will go over more details regarding maze generation in Chapter 6.

Figure 5-15.   *The rover (gray circle) is now exploring a maze*

Save the maze as `ai_rover_remastered/worlds/catacomb.world` (and load it into your Gazebo environment).

Summary

In this chapter, we have used Xacro to simplify the rover development by using modular design. This technique gave us the ability to tack on sensors such as the LiDAR and camera. We tested the rover in both the RViz and Gazebo environments using keyboard commands. Then we showed we could control the rover without the keyboard using the Teleops ROS node program. Finally, we created a simple maze to explore with our rover. In the next chapter, we will develop a rover with an embedded controller. We also reviewed and made the first use of SLAM libraries provided with ROS. However, future chapters will allow us to refine our skills with SLAM.

EXERCISES

Exercise 5.1: How would you save the constructed maps?

Exercise 5.2: How would you integrate other sensors, such as depth cameras, into the rover?

Exercise 5.3: Why did we create a digital twin before building a real rover? (Hint: Think expense and complexity.)

Exercise 5.4: Let us assume we want to use another range sensor, such as a depth camera (RGB-D camera). How can you create a plug-in to accommodate this new sensor?

CHAPTER 6

Sense and Avoidance

In Chapter 5, we used Gazebo and RViz to simulate the AI rover's behavior. The input could be scripted or arrive through human control. We also integrated a LiDAR sensor to "visualize" obstacles in the AI rover's vicinity. However, the AI rover's differential drive controller plug-in and the LiDAR sensor plug-in did not communicate with each other in Chapter 5. This lack of communication between these plug-ins led to less intelligent movement around obstacles.

We need these plug-ins to share data for intelligent navigation and trajectory control. To bridge this communication divide, we must create another plug-in with Python that sends and retrieves data from the plug-ins. This chapter will develop some simple controlling algorithms for the AI rover's differential drive controller.

We will then have an embedded, intelligent decision-making node that allows our AI rover to explore the Egyptian catacombs by sensing and avoiding obstacles, traps, and collapsed walls. This node will serve as the prototype for the cognitive deep learning engine developed in later chapters.

Objectives

The following are the objectives required for successful completion of this chapter:

- Make the AI rover detect and avoid an obstacle.

- Develop the first inertial navigation system.

© David Allen Blubaugh, Steven D. Harbour, Benjamin Sears, Michael J. Findler 2022
D. A. Blubaugh et al., *Intelligent Autonomous Drones with Cognitive Deep Learning*,
https://doi.org/10.1007/978-1-4842-6803-2_6

- Learn rover mission handling with Python.

- Develop the first Python script for obstacle avoidance.

- Understand how to upload and download Gazebo missions.

- Verify and validate the first land rover controller within Gazebo simulations.

- Discuss and review why deep learning is superior to standard state machine controller designs.

- Summary, exercises, and hints.

Understanding Coordinate Systems

Before we try to program the rover to sense its environment, we need to understand how to identify locations, speeds, and velocities. For instance, while you are reading this book, you are sitting still. Your speed is zero kph, right? Obviously!

Not so fast! From your point of view (POV), you are sitting still. But what about the alien hovering near the moon? They see you briskly speeding along at 1,673 kph because of the Earth's rotation.[1] Or the aliens hovering near the sun that see you traveling at a whopping 107,000 kph.[2]

What may feel strange is that ALL of these measurements of your speed are correct. It all depends on your POV. These POVs represent different coordinate systems. Another way of thinking about this in mathematical terms is where is the origin, O(0,0,0), of your "universe." The stationary

---

[1] Calculated at the equator: $V_{eq} = C_{eq} \div T_{eq}$, i.e., your velocity at the equator is equal to the circumference at the equator divided by the time to rotate around the circumference.

[2] Speed around the sun: $V_{orb} = 2\pi r / T_{orb}$, i.e., the velocity is the time it takes an object to complete an orbit. Rotational speed is ignored.

reader has an internal concept of everything relative to its current location. In other words, their origin moves with them. The alien near the moon sets its origin at the moon, with one face always facing the rotating Earth. Hence, the rotational speed is observed. Finally, the alien near the sun places its origin at the sun and it sees the Earth rotating and orbiting the sun. The orbital speed dominates the observation. Keen observers will notice this is a simplified explanation of Einstein's theory of relativity.

So, when solving a problem in 3D space, we find that solutions to our questions will change depending on our POV. We refer to these POVs as coordinate systems. It behooves us to use the coordinate system that solves our problem the easiest and translates into the other systems. Describing our speed and turning is easiest to understand from our internal coordinate system. Creating a map or following a path is easiest to describe from a local coordinate system. And, finally, creating our simulated world is easiest to describe in a global system. Doing the math to translate answers into the other coordinate systems is handled automatically by Gazebo.

Modeling the AI Rover World

Retrieve all of the Xacro, URDF, launch, and Python script files developed in Chapter 5. Use the same AI rover model, the differential drive controller, and the LiDAR sensor plug-in. The LiDAR sensor plug-in determines the distance between an obstacle and the AI rover. This is essential to identify the nearest obstacles during exploration. What we expect to find is that this is not enough information to navigate a maze. The rover will become trapped. Using a camera *and* LiDAR sensors will supply enough information to navigate a maze.

The AI rover is composed of the following components:

- Chassis and caster wheel: link type element

- Right and left wheels: both link type element

- Right and left wheel joints: both joint type element

- LiDAR sensor package and plug-in

Organizing the Project

Add the following highlighted sub-directories to the project:

```
└──────── sim_gazebo_rviz_ws
└──── src
    └──────── ai_rover_remastered_description
        ├──── CMakeLists.txt
        └──── package.xml
        └──── urdf
                    └─ai_rover_remastered.xacro
                    └─ai_rover_remastered_plug-ins.xacro
            └──── launch
                    └─ai_rover_rviz_gazebo.launch
                    └─ai_rover_teleop.launch
            └──── config
    └──── ai_rover_worlds
            ├──── CmakeLists.txt
            ├──── launch
            ├──── package.xml
            └──── worlds
```

To generate the ai_rover_worlds ROS package directory, execute the following Linux terminal commands:

```
$ cd ~/sim_gazebo_rviz_ws/src
$ catkin_create_pkg ai_rover_worlds
$ cd ~/sim_gazebo_rviz_ws/src/ai_rover_worlds
```

```
$ cd mkdir -p launch worlds
$ cd ../../
$ catkin_make
$ source devel/setup.sh
```

This directory organization efficiently manages the complex obstacles. The ai_rover_worlds directory includes the URDF files describing the mazes and obstacles for each generated world. The launch folder has the XML files that initiate the world and obstacles required by Gazebo. The worlds folder has the XML files that describe the world to be generated in Gazebo.

Modeling the Catacombs (Simplified)

We can now create our first AI rover's exploration world layout for the Egyptian catacombs by using the Gazebo editor, as we briefly reviewed in both Chapters 4 and 5. We will now create a simple one-room maze with a doorway and a cylindrical object (Figure 6-1). Refer to Chapter 5 about operating the Gazebo editor.

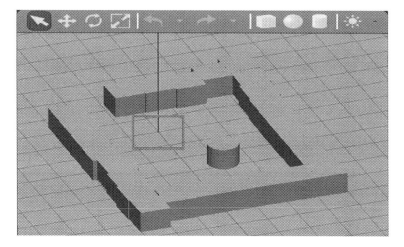

Figure 6-1. *First design of Egyptian catacomb with the column*

Now that we have created our first partially enclosed Egyptian catacomb with a supporting column, we should save this 3D design in the following directory: `~/sim_gazebo_rviz_ws/src/ai_rover_worlds/worlds`. The name for this file is `ai_rover_catacombs.world`. We should save this design by using the Ctrl+Shift+S option. The design save option allows us to store this initial design of the Egyptian catacombs without inadvertently causing the Gazebo simulator to crash. The Gazebo environment, depending on your particular computer hardware characteristics, has the potential to crash during a save. We can also see that the simulated granite blocks are misaligned within this simple catacomb design. This construction was typical for some of the earliest Ancient Egyptian Old Kingdom funerary complexes (circa 3100 B.C.). These examples of construction were especially true for those of the Ancient Egyptians who were not of the nobility class. The misaligned granite blocks will also allow us to test whether the AI rover's LiDAR and camera sensors can determine and identify uneven or misaligned surfaces' boundaries during its simulation run. We will also see the phenomenon of "blind spots" that the AI rover may not be able to locate with its sensor arrangement. Once we have this design saved in the correct directory, we must then launch it from ROS. We can launch this world within ROS, at first, without a launch file. We can do this by executing the following Linux terminal commands in two separate windows within the Terminator terminal command:

```
$ roscore                        (First Terminator Terminal)
$ rosrun gazebo_ros gazebo ~/    sim_gazebo_rviz_ws/
src/ai_rover_worlds/worlds/ai_rover_catacomb.world
(Second Terminator Terminal)
```

Note   Launch this initial design by using the Linux terminal commands. It appears to be the best method to test ROS's Gazebo interaction. However, this is still a guess on our part. We modify the launch file from Chapter 5 to generate our "catacombs" and AI rover together. It is a simple exercise to spawn additional AI rovers.

Once we have executed these Linux terminal commands, we should see the same display found in Figure 6-2. We do this to test, verify, and validate that the catacomb design can be successfully opened and operated by ROS. We will then proceed to modify our original Gazebo and RViz launch file from Chapter 5. The name for this original launch file is ai_rover_remastered_gazebo.launch. This launch file spawns and then activates a single AI rover within an empty world. We will now modify this launch file so the AI rover spawns at the Egyptian catacombs' origin coordinates developed in Figure 6-2. The origin coordinates have been highlighted in the red box in Figure 6-2, which intersects the red (x), green (y), and blue (z) lines. As such, the AI rover will spawn in the red box with the Ancient Egyptian catacombs generated around it. We must also be certain that the catacomb world never overlaps the spawn location of the AI rover. This action would cause the Gazebo simulation to terminate. Listing 6-1 is the source code listing of this modified launch file for the AI rover model and the generated catacombs.

Listing 6-1. Gazebo Launch File, Version 6.0

```
1 <launch>
2
3    <!--Robot Description from URDF-->
4    <param name="robot_description"
5        command="$(find xacro)/xacro.py '$(find
```

```
 6                    ai_rover_remastered_description)/urdf/ai_
                      rover_remastered.xacro'"/>
 7          <node name="robot_state_publisher"
                    pkg="robot_state_publisher"
 8                      type="robot_state_publisher"/>
 9

10      <node name="joint_state_publisher"
                    pkg="joint_state_publisher"
11                      type="joint_state_publisher"/>
12

13      <!--RViz-->
14      <node name="rviz" pkg="rviz" type="rviz"
                 required="true"/>
15

16      <!--Gazebo empty world launch file-->
17      <include file="$(find
                   gazebo_ros)/launch/empty_world.launch">
18         <arg name="world_name"
                      value="$(find
                   ai_rover_worlds)/worlds/ai_rover_catacomb.world"/>
19
20         <arg name="debug" value="false" />
21         <arg name="gui" value="true" />
22         <arg name="paused" value="false"/>
23         <arg name="use_sim_time" value="false"/>
24         <arg name="headless" value="false"/>
25         <arg name="verbose" value="true"/>
26      </include>
27

28      <!--Gazebo Simulator-->
29      <node name="spawn_model" pkg="gazebo_ros"
                    type="spawn_model"
30                      args="-urdf -param
```

```
                 robot_description -model ai_rover_remastered"
31                      output="screen"/>
32
33 </launch>
```

There is only one modification to the original launch file from Chapter 5: to add the `<arg name="world_name" value="$(find ai_rover_worlds)/worlds/ai_rover_catacomb.world"/>` line of XML source code following the line of source code that first launches the empty world. This world acts similar to a clean canvas, after which we generate the simulated world of the AI rover exploring the Egyptian catacombs. Therefore, we add this line to find and initiate the `ai_rover_catacomb.world` in its correct directory. We will now rename the modified version of the launch file from Chapter 5 as `ai_rover_cat.launch.` The following will be the terminal Linux commands required to execute this launch file with both the AI rover and its operational LiDAR system and the Egyptian catacombs generated together:

```
$ cd ~/sim_gazebo_rviz_ws/src
$ roslaunch ai_rover_worlds ai_rover_cat.launch
```

We should now have two simulators: 1) The RViz simulator with the AI rover at the origin; and 2) the Gazebo environment with the Egyptian catacombs, AI rover (red box), and the LiDAR sensor sweep analysis (blue area). These generated items are shown in Figure 6-2.

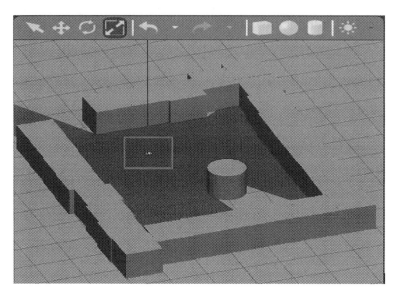

Figure 6-2. *Generated Egyptian catacombs and AI rover*

Use the Teleops program to test the AI rover's keyboard control
commands within the catacombs (Chapter 5). Reusing the Teleops test
and verification program reduces development time. We need to see if the
Teleops program still controls the AI rover prototype to circumvent the wall
boundaries and the column in the catacombs. Execute the following Linux
terminal command in a secondary shell window:

```
$ roslaunch ai_rover_remastered_description ai_rover_
teleop.launch
```

If all has worked correctly, you can now control the AI rover by
keyboard commands. As such, you should now see something similar to
Figure 6-3.

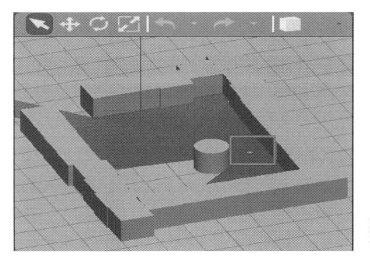

Figure 6-3. *Manually controlled AI rover exploring catacombs*

Now that we can navigate the AI rover in Gazebo, let us verify it works in RViz. Launch RViz by initiating `ai_rover.cat.launch`. Add `Laser_scan` as a topic to the Displays options found within RViz. We will now see the display shown in Figure 6-4 (see Chapter 5 for appropriate settings).

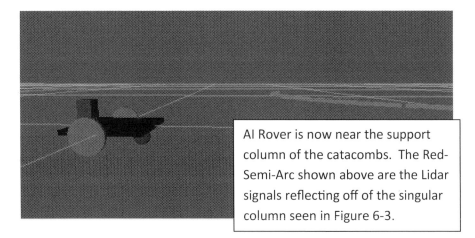

AI Rover is now near the support column of the catacombs. The Red-Semi-Arc shown above are the Lidar signals reflecting off of the singular column seen in Figure 6-3.

Figure 6-4. *RViz data display of AI rover exploring catacombs*

Note   If you receive the following error:

RLException: [ai_rover_teleop.launch] is neither a launch file in package [ai_rover_remastered_description] nor is [ai_rover_remastered_description] a launch file name. The traceback for the exception was written to the log file.

Re-execute the following Linux terminal commands (source code or launch files might need recompilation at the home directory):

```
$ cd ~/sim_gazebo_rviz_ws/
$ catkin_make
$ source devel/setup.sh
```

Next, we design, develop, test, verify, and validate the motion-planning, navigation, and map-creating script files using the simulated worlds in the ai_rover_worlds directory. We want to expand the catkin_ws directory to include intelligent control scripts, such as laser scan data extraction, to control the AI rover's differential drive. Our goal is to create the following directory tree for the catkin_ws workspace:

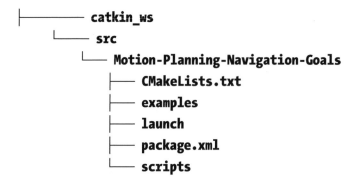

```
├────────── catkin_ws
    └────── src
        └── Motion-Planning-Navigation-Goals
            ├── CMakeLists.txt
            ├── examples
            ├── launch
            ├── package.xml
            └── scripts
```

Enter the following Linux terminal commands to create the directory tree listing for the catkin_ws workspace:

```
$ mkdir -p catkin_ws    (If catkin_ws exists, do not
recreate it.)
$ cd ~/catkin_ws/
$ mkdir -p build devel src  (Do not remake if already exist.)
$ cd ~/catkin_ws/src
```

Create the catkin ROS package Motion_Planning_Navigation_Goals using the following Linux terminal commands:

```
$catkin_create_pkg Motion_Planning_Navigation_Goals rospy std_
msgs geometry_msgs sensor_msgs
$ cd ..
$ catkin_make
$ source devel/setup.sh
```

Let us quickly review the ROS package dependencies. The rospy package is a Python client library for ROS, allowing programmers to access the ROS control libraries. However, the rospy library is not computationally efficient. Instead, the library allows the programmer to quickly and efficiently develop ROS applications.

The std_msgs dependency develops standard publisher/subscriber ROS messages, including common data types. These data types include float and double built-in data types and user-defined data types, such as the deep learning neural network or multi-arrays for geospatial processing.

The geometry_msgs dependency provides messages for geometric elements (points, vectors, and poses). The geometry_msgs offer standard functions, such as twist. The twist function describes an element's linear and angular velocities for both subscribers and publishers. The differential drive plug-in uses these velocities. We need this geometry_msgs dependency to communicate the Python autonomous program's linear and angular velocities to control the rover's differential drive.

The sensor_msgs dependency provides sensor messages, such as cameras and range-finding laser scanners. This dependency allows for information exchange from both the publishers (sensors) and subscribers (Python autonomous programs). These sensor messages influence the Python autonomous programs to vary the differential drive plug-in's linear and angular velocities. For more information on the LiDAR's laser pipeline stacks, please refer to http://wiki.ros.org/sensor_msgs.

Laser Range-Finding Filter Settings

Set the LiDAR sensor settings to the values shown in Table 6-1 in the LiDAR sensor section of the ai_rover_gazebo_plug-in.xacro file.

Table 6-1. *LiDAR-Specific Program Settings*

XML Setting	Value	Definition
<update_ rate>	20	Controls the sample rate (speed) of the laser data capture rate
<samples>	1440	How many samples for one LiDAR range sweep [minAngle..maxAngle]
<frameName>	sensor_laser	Virtual link between LiDAR sensor and chassis

Verify the program values fall within the LiDAR sensor's actual physical specifications. The Gazebo simulator creates a more realistic LiDAR sensor with these values.

After saving these values, determine if the updates were successful by executing the following in one window:

```
$ cd ~/sim_gazebo_rviz_ws/src
$ roslaunch ai_rover_worlds ai_rover_cat.launch
```

And execute the following in a second window:

```
$ cd ~/sim_gazebo_rviz_ws/src
$ roslaunch ai_rover_remastered_description ai_rover_
teleop.launch
```

Once we have executed these commands, we can control the AI rover displayed in a simulation using the keyboard. Next, we will develop a Python script code to process and display the LiDAR sensor data.

Laser Range-Finding Data

We will create our Python script within the catkin_ws workspace Motion_ Planning_Navigation_Goals ROS package. This script will retrieve data from the /ai_rover_remastered/laser_scan/scan topic. Create the following Python script:

```
$ cd ~/catkin_ws/src
$ cd Motion_Planning_Navigation_Goals
$ mkdir scripts
$ cd scripts
$ touch read_Lidar_data.py
$ chmod +x read_Lidar_data.py
$ nano read_Lidar_data.py
```

Note   The touch command creates a file within directories without launching an application, such as a word processor or a development environment. The created files are empty. To create content, the appropriate application must be launched. The chmod command makes the read_Lidar_data.py file executable via rosrun.

Enter the script found in Listing 6-2.

Listing 6-2. *read_Lidar_data.py*

```
1  #! /usr/bin/env python3
2
3  import rospy
4  from sensor_msgs.msg import LaserScan
5  def clbk_laser(msg):
6
7      # 1440/10 = 144
8      sectors = [
9          min(min(msg.ranges[0:143]), 10),
10         min(min(msg.ranges[144:287]), 10),
11         min(min(msg.ranges[288:431]), 10),
12         min(min(msg.ranges[432:575]), 10),
13         min(min(msg.ranges[576:720]), 10),
14         min(min(msg.ranges[720:863]), 10),
15         min(min(msg.ranges[864:1007]), 10),
16         min(min(msg.ranges[1008:1151]), 10),
17         min(min(msg.ranges[1152:1295]), 10),
18         min(min(msg.ranges[1296:1439]), 10)
19         ]
20
21     rospy.loginfo(sectors)
22
23 def main():
24
25     rospy.init_node('read_Lidar_data')
26     sub=rospy.Subscriber("/ai_rover_remastered/laser_scan/
       scan", LaserScan, clbk_laser)
27
```

```
28        rospy.spin()
29
30   if __name__ == '__main__':
31
32        main();
```

The clbk_laser function passes the msg argument to min(min(msg. ranges[xxxx:yyyy]), 10), which then reads the data for each of the ten LiDAR sensor sectors, guaranteeing a maximum value of 10 meters. It converts the 1,440 samples into ten aggregated sectors. Each sector returns the minimum value from the samples or caps the value at 10 meters (LiDAR max range). The minimum value is the distance to the obstacle closest to the AI rover in that sector. Each sector is 36 degrees in coverage (36° x 10 = 360°). As such, the AI rover has no "blind spots" during its continuous LiDAR sensor sweep. Because the AI rover has no blind spots, it can also travel backward if necessary. Figure 6-5 is the orientation layout of the LiDAR sensor.

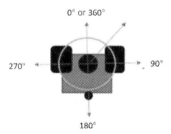

Figure 6-5. *LiDAR sensor orientation and layout*

The larger solid black circle in the center of Figure 6-5 is the LiDAR sensor. The smaller solid black circle toward the bottom of Figure 6-5 is the caster wheel. The LiDAR sweep begins at 180° and then ends at 180°. The red arrow is the propagating LiDAR laser beam.

To run our program `read_Lidar_data.py`, add the following main test driver found in lines 23–32. The source code contained within `main()` performs the following functions:

- **rospy.init_node('read_Lidar_data')**: Attaches to ROS.

- **rospy.Subscriber():** Subscribes to the `Laser_scan` and `msg_Lidar_laser` topics.

- **rospy.Publisher():** Publishes to the `ai_rover_ remastered base_controller/cmd_vel` topic for controlling the AI rover.

- **rospy.spin():** Keeps the program from shutting down on its own. The program can only be shut down by the operator.

To run the `read_Lidar_data.py rosrun` executable, reenter the following Linux terminal commands:

```
$ cd ~/catkin_ws
$ catkin_make
$ source devel/setup.sh
```

After recompiling everything, run the program with the following:

```
$ cd ~/catkin_ws/src
$ cd Motion_Planning_Navigation_Goals
$ cd scripts
$ rosrun Motion_Planning_Navigation_Goals read_
  Lidar_data.py
```

Note   If either the Gazebo AI rover catacomb program or keyboard Teleops is not running, recompile/load both launch files and try running the `read_Lidar_data.py` program again.

Your data should look like that in Figure 6-6.

[INFO] [1598597265.044337]: [0.7486084699630737, 0.8982205986976624, 2.04443025
58898926, 2.151888608932495, 3.4493050575250348, 3.078852415084839, 3.161072731
0180664, 3.0056800842285156, 0.9493757486343384, 0.7484521269798279]

LiDar sensor data array of 10 36°C intervals. Each array element is the shortest distance to an obstacle within that 36° interval. As we can see that 90° and a wall is just behind the rover. That is why we have 0.748 m distance.

Figure 6-6. *LiDAR sensor analysis at 0.748 meters*

We should perform a sanity check and visually inspect that the AI rover is currently 0.75 meters from the wall. Figure 6-7 shows our numbers make sense. Moving the AI rover a little southeast, our numbers change to those in the figure.

[INFO] [1598600815.242010]: [1.736396312713623, 2.572085380554199, 2.5696382522 58301, 3.2980737686157227, 2.0650675296783447, 2.141602039337158, 2.44478774070 73975, 2.4591922760009766, 1.7071266174316406, 1.6250174045562744]

LiDar sensor data array after moving forward. We can see the first array data number has increased from 0.75 – 1.736 m. This is about what would be expected.

Figure 6-7. LiDAR sensor analysis at 1.736 meters

We have determined that the LiDAR sensor system is operational. We conducted this small distance test to verify the Gazebo simulations are working correctly. The simulated AI rover must as closely as possible operate within the laws of physics. By performing these iterative tests of every single development of the AI rover, our odds of building a fully operational AI rover are improved. It also narrows the location we need to search when looking for the source of an error.

Use Table 6-2 to visualize and interpret the 10-element array shown in Figure 6-7. The sensor data is simplified to [INFO] [1598600815.242010] [1.76, 2.57, 2.56, 3.29, 2.06, 2.14, 2.44, 2.45, 1.71, 1.63].

Table 6-2. *LiDAR Sensor Readings from Figure 6-7, Interpreted*

Sector	Direction	Degrees of Coverage	Nearest Object
SectorA	back_right	144° to 180°	1.76 meters
SectorB	back_center_right	108° to 144°	2.57 meters
SectorC	Right	72° to 108°	2.56 meters
SectorD	front_center_right	36° to 72°	3.29 meters
SectorE	front_right	0° to 36°	2.06 meters
SectorF	front_left	324° to (360° or 0°)	2.14 meters
SectorG	front_center_left	288° to 324°	2.44 meters
SectorH	Left	252° to 288°	2.45 meters
SectorI	back_center_left	216° to 252°	1.70 meters
SectorJ	back_left	180° to 216°	1.62 meters

Each sector represents 1/10th of the LiDAR sweep (Figure 6-8). SectorA detected an object ~1.76 m away from the AI rover. There may be other objects in the sector, but none closer. The object in SectorA is located between 144° and 180° with respect to the AI rover. The object may also be in two sectors; i.e., the sensor data does not infer size. All of the other sectors can be interpreted similarly.

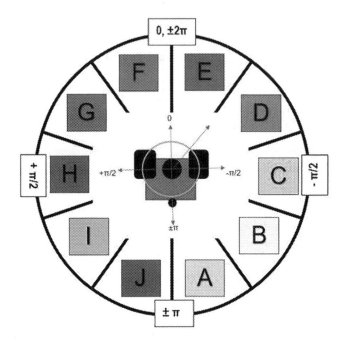

Figure 6-8. *LiDAR sensor sector orientation and layout*

Obstacle Sense-and-Avoidance

This section bridges the information presented in Chapters 1 through 5 (manual control) with the remaining chapters (intelligent autonomous control). We now take our first steps toward developing an autonomous rover; i.e., it makes its own decisions through the maze. It achieves this through a simple control algorithm. The control algorithm sends commands to the actuators/motors.

Organize the control algorithm folder directories as follows:

```
$ cd ~/sim_gazebo_rviz_ws/
$ catkin_make
$ source devel/setup.sh
$ cd ~/catkin_ws/src
```

```
$ cd Motion_Planning_Navigation_Goals
$ cd scripts
$ touch sense_avoid_obstacle.py
$ chmod +x sense_avoid_obstacle.py
$ nano sense_avoid_obstacle.py
```

This creates an empty sense_avoid_obstacle.py file in the scripts directory. The following is the tree structure of listing files and folder directories:

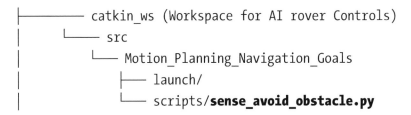

```
├──────── catkin_ws (Workspace for AI rover Controls)
│    └──── src
│         └──── Motion_Planning_Navigation_Goals
│              ├──── launch/
│              └──── scripts/sense_avoid_obstacle.py
```

Now that we have an understanding of our tree-structure organization, we must return our attention to the source code contained within the sense_avoid_obstacle.py ROS program node. Listing 6-3 provides this source code.

Listing 6-3. *sense_avoid_obstacle.py Version 6.0. No Artificial Intelligence Yet!*

```
1 #! /usr/bin/env python
2 # -*- coding: utf-8 -*-
3
4 import rospy
5 from sensor_msgs.msg import LaserScan
6 from geometry_msgs.msg import Twist
7
8 pub = None
9
```

```
10 # 1440/10 = 144 Full rotation of coverage from +/-Ï€ for
   the LiDAR sensor sweep
11 # Transform â€˜sectorsâ€™ array into a Python dictionary.
12
13 def msg_Lidar_laser(msg):
14     sectors = {
15     'sectorA': min(min(msg.ranges[0:143]),    10),
16     'sectorB': min(min(msg.ranges[144:287]), 10),
17     'sectorC': min(min(msg.ranges[288:431]), 10),
18     'sectorD': min(min(msg.ranges[432:575]), 10),
19     'sectorE': min(min(msg.ranges[576:719]), 10),
20     'sectorF': min(min(msg.ranges[720:863]), 10),
21     'sectorG': min(min(msg.ranges[864:1007]), 10),
22     'sectorH': min(min(msg.ranges[1008:1151]),10),
23     'sectorI': min(min(msg.ranges[1152:1295]),10),
24     'sectorJ': min(min(msg.ranges[1296:1439]),10),
25     }
26     rospy.loginfo(sectors);
27     sense_and_avoid(sectors);
28
29 # Using encoded sensor data to make turning decisions.
30 # There are five different possible decisions: Forward,
   turn soft left, turn hard
31 # left, turn soft right, turn hard right, and turn around
   180 degrees aroundâ€¦
32
33 def switch_movement_options(argument):
34     def forward():
35         return 0.5, 0.0;
36
37     def turn_soft_left():
```

```
38          return 0.0, 0.70;
39
40      def turn_hard_left():
41          return 0.0, 0.90;
42
43      def turn_soft_right():
44          return 0.0, -0.70;
45
46      def turn_hard_right():
47          return 0.0, -0.90;
48
49      def backward_turn_around():
50          return 0.0, 3.14159;
51
52      switch_function_dir = {
53
54          0b0000: forward(),
55          0b0001: turn_soft_left(),
56          0b0010: turn_soft_left(),
57          0b0011: turn_soft_left(),
58          0b0100: turn_soft_right(),
59          0b0101: turn_hard_left(),
60          0b0110: turn_hard_left(),
61          0b0111: turn_hard_left(),
62          0b1000: turn_soft_right(),
63          0b1001: forward(),
64          0b1010: turn_hard_right(),
65          0b1011: turn_hard_left(),
66          0b1100: turn_hard_right(),
67          0b1101: turn_hard_right(),
68          0b1110: turn_hard_right(),
```

```
69              0b1111: backward_turn_around()
70              }
71
72      return switch_function_dir.get(argument, "Invalid Move
        Option")
73
74 def sense_and_avoid(sectors):
75      msg = Twist();
76      linear_vel_x = 0;
77      spin_angular_vel_z = 0;
78      current_state_description = 'Program Start';
79
80      # encode forward looking sensor data
81      bitVar = 0b0000;
82
83      # bitVar | 0b1000 We have an obstacle detected in
        front_center_left
84      if sectors['sectorG'] < 1:
85          bitVar = bitVar | 8;
86
87      # bitVar | 0b0100 We have an obstacle detected in
        front_left
88      if sectors['sectorF'] < 1:
89          bitVar = bitVar | 4;
90
91      # bitVar | 0b0010 We have an obstacle detected in
        front_right
92      if sectors['sectorE'] < 1:
93          bitVar = bitVar | 2;
94
95      # bitVar | 0b0001 We have an obstacle detected in
        front_center_right
```

```
96       if sectors['sectorD'] < 1:
97           bitVar = bitVar | 1;
98
99       # SET ALL VALUES TO BE SENT TO THE DIFFERENTIAL DRIVE
         VIA TWIST
100      linear_vel_x, spin_angular_vel_z = switch_movement_
         options(bitVar);
101
102      rospy.loginfo(current_state_description);
103      msg.linear.x = linear_vel_x;
104      msg.angular.z = spin_angular_vel_z;
105      # publish the values of linear and angular velocity to
         diff drive plug-in
106      pub.publish(msg);
107
108 def main():
109      global pub;
110
111      rospy.init_node('sense_avoid_obstacle');
112      sub = rospy.Subscriber('/ai_rover_remastered/laser_
         scan/scan', LaserScan, msg_Lidar_laser);
113
114      pub = rospy.Publisher('ai_rover_remastered/base_
         controller/cmd_vel', Twist, queue_size=10);
115
116      rospy.spin();
117
118 if __name__ == '__main__':
119      main();
```

Source Code Analysis

There are ten functions defined within this program, including six for wheel movement. The last function (`main` 108–119) is similar to the other versions of the `main` previously discussed in the `read_Lidar_data.py` program. We will ignore lines of code that are repeated from previous iterations.

We will now define the following contents of the `main` function.

- Lines 13–25. Retrieves data from the LiDAR sensor arcs, capping them at a maximum of 10 meters. For instance, `'sectorA': min(min(msg.ranges[0:143]), 10)` grabs the smallest value from all 144 data samples from the first arc, SectorA, and then caps that at 10. So, if all 144 data samples are greater than 10, SectorA will be assigned the value of 10. For more information about the interpretation of the LiDAR data, refer to the "Interpreting the LiDAR" section in this chapter. Additionally, review the `read_Lidar_data.py` program in "Laser Range Finding Data."

- Lines 26–27. Saves values to log file and then calls `sense_and_avoid` (line 94)

- Lines 28–70. These similar-looking functions just set the linear velocity in the x-axis (how fast we are going forward) and the angular velocity in the z-axis (how fast we turn). For instance, if there are objects in the right quadrant, we can turn left at a relatively slow speed. If there are objects directly in front of us, we want to quickly turn in another direction.

Note 1   When we turn, there is no forward momentum. We either move forward or we turn, not both. This makes coding a lot easier.

Note 2   The turning command is how fast we turn in a given direction, not a target angle. The sensor readings happen approximately 10 times per second. Therefore, if we tell the rover to turn at a rate of 45° per second, the rover will have turned only 4.5° when a new sensor reading takes place. This new reading overrides the previous Twist() command: 1) This is why we use "soft" and "hard" turns. 2) This can and will cause an infinite loop, where the rover turns back and forth. We will talk about and review a control systems solution to this issue in the Chapter 5 "Summary" section.

Taking the next section of code out of order:

- Lines 74–99. This section encodes the four sensors. Initially, we assume there are no close objects (set bitVar to 0b0000). If an object is close (sector value < 1m), we set the appropriate bit in bitVar to 1. As illustrated in Figure 6-9, if there is an object in Sector E, we "flip" the SectorE bit (0b00x0), where the bit in the second location is changed. This is the "2's" place, hence the bit-or bitVar = bitVar | 2; flips the "2's" place (See Figure 6-9). For more information regarding the sense_and_avoid function please refer to the "Sensing and Avoiding Obstacles" section.

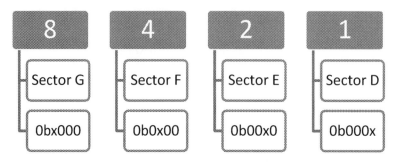

Figure 6-9. *Bit location encoding for each forward sector*

Note   This is not our first iteration of this piece of code. After our first iteration, we recognized that the sixteen different cases represented by the four sensors (D, E, F, and G) only resulted in six possible outcomes (Forward, Soft_Left, Hard_Left, Hard_Right, Soft_Right, and Retreat). Furthermore, encoding the four sectors into a single 4-bit variable sped up the code tremendously. This code rewriting is called *refactoring*.

- Lines 52–72. With the encoded bitVar, we can do a quick "dictionary" lookup of the appropriate rover movement. The movement decision is based on the location of nearby objects encoded in "arguments." Line 125 passes the encoded bitVar to switch_movement_options(arguments), and this chooses the "correct" direction to move.

Interpreting the LiDAR Sensor Data

The callback function, `msg_Lidar_laser`, handles information and messages from the LiDAR sensor for the subscriber defined in the main function (line 112). The `msg_Lidar_laser` function partitions the LiDAR sensor sweep messages into ten equally spaced 36-degree sectors for the AI rover. The LiDAR sends 1,440 laser pulses every revolution. Every pulse is a ray originating from the LiDAR and going to infinity UNLESS the ray encounters an obstruction. The pulse reflects the LiDAR, and, based on the return time, the distance to the obstruction is calculated. All 1,440 data points are sent by the LiDAR as an event that the `msg_Lidar_laser` is listening for and are processed into the ten equally spaced sectors.

The purpose of having ten equally spaced 36-degree sectors is to cover one entire revolution around the AI rover. This allows the AI rover to "see" and travel in any required direction, including forward, left, right, and backward. Traveling in any direction allows the AI rover to navigate and avoid obstacles and cave-ins within the Egyptian catacombs.

Sensing and Avoiding Obstacles

The `sense_and_avoid` function is a simple finite state machine (FSM) implementing simplistic behavior (see Figure 6-10).

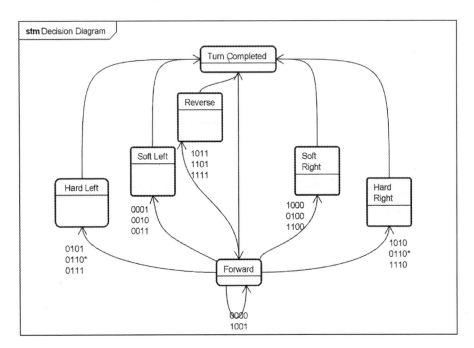

Figure 6-10. *Decision finite state machine showing different turning decisions*

When developing a complex system, such as an intelligent drone, approach the solution in small, incremental steps. To simplify our decision-making choices about turning, we are only going to use the front four sectors' data; i.e., sectors D, E, F, and G. This sensor data is used to determine if objects exist in our path that are less than one meter away. These four sectors make sixteen combinations that determine the direction the rover will go. There are only six possible directions: forward, soft left, soft right, hard left, hard right, and turn around (see Figure 6-11). (If the "front six" sectors were used for the decision-making, we would have 64 combinations! Much more complex!)

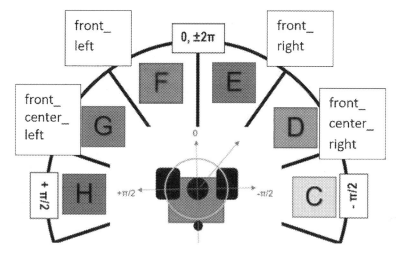

Figure 6-11. *Forward-facing sectors are used to decide the next move*

Each of the four forward-facing sectors returns the distance of the nearest object in that sector. As can be seen in Table 6-3, we encode the results in bitVar. For example, SectorG might return the value 10. That means the object is 10 meters away and not of immediate concern. Assume that SectorF and SectorE are similarly large numbers and no danger. Let us say that SectorD returns 0.9. This object is less than 1m away and a possible hazard that we should avoid. Testing whether each returned value is less than one (i.e., SectorG < 1) returns a true or a false. In this example, values returned by SectorG, SectorF, and SectorE are large, and the comparison will return false. In contrast, SectorD's comparison will return true. The four comparisons are false, false, false, true, which can be encoded as bitVar = 0001. Interpreting Table 6-3: If a cell is blank, the nearest object in that direction is far away and of no immediate danger. If the cell has "< 1m," then an object is very close and should be avoided.

Table 6-3. *Decision Table Using the Four Forward Sectors to Choose One of Six Decisions. Blank Cell Indicates No Nearby Object Detected in Those Sectors*

Case	bitVar	SectorG	SectorF	SectorE	SectorD	Decision
0	0000	No nearby objects *				Forward
1	0001				< 1 m	Soft Left
2	0010			< 1 m		Soft Left
3	0011			< 1 m	< 1 m	Soft Left
4	0100		< 1 m			Soft Right
5	0101		< 1 m		< 1 m	Hard Left
6	0110		< 1 m	< 1 m		Hard Left
7	0111		< 1 m	< 1 m	< 1 m	Hard Left
8	1000	< 1 m				Soft Right
9	1001	< 1 m			< 1 m	Forward
10	1010	< 1 m		< 1 m		Hard Right
11	1011	< 1 m		< 1 m	< 1 m	Turn Around
12	1100	< 1 m	< 1 m			Soft Right
13	1101	< 1 m	< 1 m		< 1 m	Turn Around
14	1110	< 1 m	< 1 m	< 1 m		Hard Right
15	1111	< 1 m	< 1 m	< 1 m	< 1 m	Turn Around

Formally, the rover's speed and direction are called the linear velocity and angular rotation in the rover's internal coordinate system. The linear velocity is expressed along the x-axis (red), and the angular rotation is about the z-axis (blue), as shown in Figure 6-12. Since we are using a rover and assuming flat ground at this time, the system reduces to 2D. After removing the "flat ground" assumption, our model will become more complex. But that is for a later chapter.

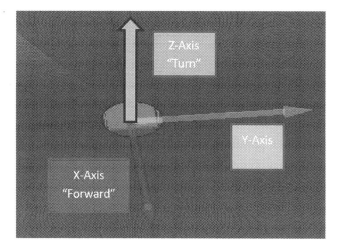

Figure 6-12. *X-, y-, and z-axes are relative to the rover*

There are two types of angular velocity: orbital and spin. For example, the Earth orbits the sun; i.e., the Earth rotates around an external point (sun). The Earth also spins about the north/south pole axis; i.e., the Earth rotates around an internal point. We use Twist() (line 96) to send the spin angular velocity to the AI rover. This causes the AI rover to rotate around the internal z-axis.

Another way of looking at orbital angular velocity (OAV) versus spin angular velocity (SAV) is the point of view. OAV is observed from outside the rover; i.e., it interacts with multiple objects in the universe. The OAV for the Earth/sun combination is 365 days, while the OAV of the moon/ Earth orbit is 28 days. OAV depends on other objects. On the other hand, SAV is an internal attribute. The SAV of the Earth is 24 hours. It does not change concerning other objects. Thus, we say the SAV is independent of the observer, whereas OAV is dependent.

When we control the rover, we are using SAV. In other words, if we want to turn right 45°, we inform ROS to turn the rover 45°. But remember, ROS uses radians instead of degrees when calculating angles. Angular velocities are expressed as radians per second, where one radian is equal to an angle at the center of the circle whose arc is equal to the radius. From our internal perspective, if we "turn left," we pass a positive value. If we "turn right," we pass a negative value. So our "turn left 45°" becomes "turn left $\pi/4$."

Executing the Avoidance Code

Execute the following Linux terminal commands in one opened terminal:

```
$ cd ~/catkin_ws/
$ catkin_make
$ source devel/setup.sh
$ cd ~/sim_gazebo_rviz_ws/
$ catkin_make
$ source devel/setup.sh
$ cd ~/sim_gazebo_rviz_ws/
$ roslaunch ai_rover_worlds ai_rover_cat.launch
```

Then, execute the following Linux terminal commands in a second terminal:

```
$ cd ~/catkin_ws/
$ rosrun Motion_Planning_Navigation_Goals sense_avoid_
obstacle.py
```

Our rover now avoids simple obstacles, such as the support column. But this avoidance algorithm is not very smart! It only uses four specific sectors, and each avoidance decision was explicitly chosen by the programmer. What if the programmer did not understand all the possible situations and all the possible decisions? What if the actual world is so sophisticated that four sectors do not supply enough information to always make the correct decision?

Using an artificial intelligence deep learning system (DLS) will allow the AI rover to have evolutionary and adaptive behaviors. The use of these advanced methods eliminates the programmer's need to explicitly program all of these behaviors and decisions with a finite-state machine (FSM). To achieve these behaviors we replace the FSM with a DLS. For example, in Figure 6-13, the FSM-based rover may get stuck in the corner because of the collapsed wall. The DLS AI rover should be able to intelligently escape the collapsed wall. We will add the DLS intelligence in the next chapter.

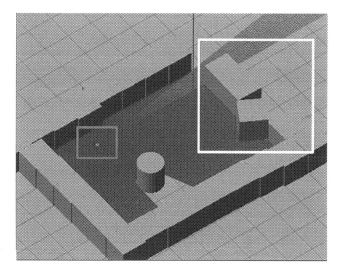

Figure 6-13. *Scenario with AI rover (red) and collapsed wall (yellow)*

Summary

The purpose of this chapter is to convey to the reader that the development of simple control is possible for the first generation of the AI rover. This chapter also exposes the severe limits of finite state machines in the creation of ever more sophisticated robots, thereby necessitating the creation of deep learning, reinforcement deep learning, Bayesian learning, and ultimately artificial cognitive intelligence–based controllers that will allow a robot to explore an unknown environment, such as the Egyptian catacombs. We should also point out to the reader that we have not used any type of probability or statistical analysis on the sensor data to determine sensor noise or error. We will revisit this topic in the chapters on Bayesian deep learning and statistical analysis for sensor noise and uncertainty. We will talk in the next chapter about the development of the first navigational and mapping techniques based on SLAM (Simultaneous Localization and Mapping). The discussion of navigation and mapping will also require the use of control system theory to allow us to understand why it is important to have two main control systems to control the angle heading of the AI rover itself and the direction of the AI rover's goal position. The next chapter will be a small but critical step in the development of an autonomous rover.

EXTRA CREDIT

1. During the AI rover/catacomb simulations, you must have noticed that if the AI rover collides with a granite block it can push it to one side. In reality, this would never happen due to the AI rover's small size, mass, and speed. How would you alter the URDF XML files to allow the granite blocks and supporting column found in the Egyptian catacombs to have the

same simulated mass and moment of inertia as a real-world granite block and column with the same dimensions? (Look to Chapters 4 and 5 on modifying the URDF source code.)

2. Is the AI rover still able to push the granite blocks to one side if it collides with a block with a real-world mass and moment of inertia?

3. We know that we currently have 16 combinations for `bitVar` and for the four front sectors that we have used (D, E, F, and G) in `sense_avoid_obstacle.py`. How many combinations for `bitVar` would we have if we added both the left (SectorH) and the right (SectorC) LiDAR sectors? Please list all of the combinations of `bitVar` by using Sectors C, D, E, F, G, and H.

4. We can see that we are only using the four front LiDAR sectors. What gains could we make if we were to incorporate both the left (SectorH) and the right (SectorC) LiDAR sectors into `sense_avoid_obstacles.py`? How would we go about reprogramming this particular Python ROS node program?

5. We are currently using ten sectors for the LiDAR sensor array sweep? Is this the only number of LiDAR sectors we could use? If not, then what would be the ideal number of LiDAR sensor sweep sectors?

6. How do we modify the source code for the simulated LiDAR sensor range from 10 meters to 30 meters?

CHAPTER 7

Navigation, SLAM, and Goals

In this chapter, our intrepid rover will explore the Egyptian catacombs. Any autonomous rover needs to know where it is in its environment; i.e., its current position. That way, the rover can travel from one location to the next without user input. To do this, the rover must construct an internal map of its environment as it travels. This chapter introduces the simultaneous localization and mapping (SLAM) algorithm. SLAM generates maps from navigation data, such as the odometry data from the rover's wheel encoders and the laser range data from the LiDAR sensor.

Objectives

The following are the objectives required for successful completion of this chapter:

- Mission types

- Tracking local and global location in an environment (odometry)

- Correcting our travel (control theory)

- Simultaneous Localization and Mapping (SLAM)

© David Allen Blubaugh, Steven D. Harbour, Benjamin Sears, Michael J. Findler 2022
D. A. Blubaugh et al., *Intelligent Autonomous Drones with Cognitive Deep Learning*,
https://doi.org/10.1007/978-1-4842-6803-2_7

Overview

To create an autonomous rover, the rover must know its position and orientation in the environment.

Mission Types

There exist three types of rover missions: complete coverage, goal directed, and exploratory. We will discuss **complete coverage** missions later. The **goal-directed** mission goes from Point A (initial) to Point G (goal). The goal can be a simple straight-line path (A connects directly to G) or a segmented path that traverses through other waypoints (A→B→C→...→G). The intermediate waypoints can be preplanned or discovered autonomously.

The **exploratory** mission discovers features of the environment and places them on internal local and global maps. These features can be walls, doorways, objects, holes in the ground, etc. Creating this internal global map is necessary so as to plan future missions with optimal paths. An exploratory mission can be one of three sub-types: one-point or two-point missions, and resource discovery.

- The one-point mission gives the rover a starting position from which to "map" the entire environment and to which it should return. This mission type would be in a new environment, such as moving to a new house and randomly exploring the entire neighborhood.

- The two-point mission provides the rover with its starting and final or goal positions. We can assume the final position is in an unexplored location on the intermediate map. The two-point mission does not have to explore the unknown environment thoroughly, just enough to get to the final goal. After moving into our new home, we know the address of the school we will attend, so we explore the route to learn about landmarks. Maybe there are some road closures, etc.

- Our third—exploratory—mission is a one-point mission with a goal resource. Assuming no GPS, we might explore to find a pizza shop when we move into our new house. We do not know where the pizza shop is. We are just exploring the neighborhood until we find one.

Odometry

Odometry estimates the change in position based on wheel sensor data. The odometry is calculated using the (simulated) rover's wheel encoder. Think of putting a playing card on your bicycle as a kid as an analogy. Each time the card struck a spoke, a click was heard. If we know the original position of the bicycle and how many clicks were generated, we could calculate the new location of the bicycle (assuming the trip was a straight line). We know this because we know how many spokes (18) the wheel has and the circumference (36 inches). So, each click translates to "move forward two inches."

Furthermore, we can calculate the bicycle's speed by the frequency of the clicks. Of course, the assumption that the bicycle is going in a straight line is relatively strong, and this also does not take into account any skidding or sliding, etc. The odometry gives us an approximation of our current position! The shorter the trip, the better the approximation.

This position approximation process is how the wheel encoder works. The rover's odometry is calculated by counting each "click" of the wheels to approximate the rover's speed, direction, and location. The distance-traveled calculation is accurate as long as the simulated rover travels in a straight line (no turns). A more sophisticated odometry algorithm combines linear (forward and backward) and angular (turning left or right) acceleration to calculate the rover's velocity and position. We can also use the inertial measurement unit (IMU) sensor to make more accurate calculations. Unfortunately, the calculations will still become increasingly erroneous over time and distance due to round-off errors, and we cannot correct this without updating our location from external sources.

Navigation, SLAM, and pathfinding algorithms use odometry's calculated position. The position is its location and its orientation, known as its **pose**. The degrees of freedom (DoF) is the number of dimensions necessary to describe the complete pose of our rover. For example, an airplane has six DoF because it can move in three dimensions <x,y,z> and rotate in three dimensions <r,p,y>, where r,p,y stands for roll, pitch, and yaw. Similarly, in the rover's URDF Xacro files, the initial position and orientation six DoF is given as follows:

```
<xacro:property name="left_sensor">
  <origin xyz="0.3 0 0" rpy="0 0 0" />
</xacro:property>
```

Finally, odometry is used in both local and global maps. The **local** map tells the rover its location based on the starting position. The **global** map places our rover within the environment for visualization by an external viewer such as the RViz environment.

Rover's Local Navigation

Our rover can travel from one location to the next in an environment, handling obstacles blocking its path based on the odometry calculations. As mentioned earlier, there are two coordinate systems for the rover's navigation: local and global. The local navigation is the internal understanding of the rover that can sense nearby objects. The rover's local sensing capability means the rover will know that it started at location (0,0) facing north; it is currently at (1,3) facing east and senses an object at (1,5). We will explore global navigation later.

Furthermore, if the rover keeps moving east, it may collide with an obstacle—it could be a wall or a stand-alone object. The local goal is to avoid the obstacle. To avoid the obstacle, we can 1) stop, 2) turn left/right, and 3) move forward into an unblocked area, OR we can turn as we move into an unblocked area. The first solution is easiest to implement, but it slows our rover's exploration. Conveniently, Robotic Operating System (ROS) has plug-ins to assist in controlling the rover, so we do not have to understand the physics.

Rover's Global Navigation

Let us assume that we are on a two-point exploratory mission. Global navigation is finding a path between two points located somewhere in the environment. Finding the route between these two points requires generating a global map of the domain. As the rover explores, we use odometry to calculate its location. We simply convert from the local coordinate system to the global coordinate system and draw our rover avatar at that location. Furthermore, sensor data populates the map with object outlines, such as walls and obstacles. After creating this map, future goal-directed missions can preplan the shortest path between the two points.

Hopefully, it is evident that errors in calculating the rover's location from the odometry can have significant repercussions on the global map rendering. Knowing where the rover is in the environment allows us to create accurate maps on exploratory missions and optimal paths on goal-directed, planned missions.

Odometry can be used for short durations with no real consequences to the rover's position and pose. However, odometry's uncertainty in determining the precise location and pose of the rover can be bothersome. This uncertainty is especially problematic when exploring in an uncertain environment for a long time. To help reduce these issues of uncertainty with missions of longer duration, we update our global maps on every mission. This reinforcement of object location can boost confidence in the global map's construction. Furthermore, an object's permanent location lets us assign position tags. To reset the odometry settings, we can use the object's permanent position tag as an external "verified" location.

Getting the Rover Heading (Orientation)

To develop a more adaptable sense-and-avoid algorithm, we need to know our rover's direction. To achieve this, the /odom topic provides our AI rover's heading sensor data to the Twist() function. Recall that Twist() gets the new desired heading and rotates the rover iteratively from its current heading to its desired heading. Then Twist() passes the data through the /cmd_vel topic to control the angular velocities of each wheel. As mentioned in Chapter 6, these controls allow the AI rover to escape tight corners and enclosed areas. Now the AI rover can avoid obstacles without external steering.

Note   We have started to create a rover that is autonomous—i.e., it can steer itself without human intervention. The self-driving appears "intelligent" to an outside observer, and this means the rover has just evolved to an artificially intelligent rover, or AI rover. Henceforth, *rover* and *AI rover* are synonymous.

Create a rotateRobotOdom.py script in the catkin_ws/src directory. This script retrieves orientation data from the /odom topic (Listing 7-1). This data lets the rover know what its heading is.

Listing 7-1. rotateRobotOdom.py

```python
1 #!/usr/bin/env python3
2 import rospy
3 from nav_msgs.msg import Odometry
4 from tf.transformations import
  euler_from_quaternion, quaternion_from_euler
5
6 roll = pitch = yaw = 0.0; # initially point North
7
8 def get_rotation (msg):
9     global roll, pitch, yaw;
10    orientation_q = msg.pose.pose.orientation;
11      orientation_list = [orientation_q.x,
          orientation_q.y,
          orientation_q.z,
          orientation_q.w];
12     (roll, pitch, yaw) =
    euler_from_quaternion(orientation_list);
13     print yaw;
14
15 rospy.init_node('rotateRobotOdom');
16
17 sub = rospy.Subscriber(
  'ai_rover_remastered/base_controller/odom',
  Odometry, get_rotation);
18
19 r = rospy.Rate(1); # publish msg every second
20 while not rospy.is_shutdown():
21        quaternion_val =
22     quaternion_from_euler (roll, pitch, yaw);
23   r.sleep();
```

Most of the script should be familiar, but a few lines need to be explained. The code updates the roll, pitch, and yaw of the AI rover as it travels and rotates. Only the yaw values change since the environment is modeled as a flat, two-dimensional surface. We will test this script by using the keyboard-controlled `teleop_twist_keyboard.py` program.

In Line 6, we face "North" by setting the roll, pitch, and yaw values to zero at the start of the mission. Lines 7–14 define the `get_rotation` function. The argument to this function is from the navigation `msg`; in quaternion coordinates: [x, y, z, w]. Do not worry about quaternion coordinates because Line 12 converts them to Euler coordinates. Line 15 registers the `rotateRobotOdom` node with ROS to subscribe and publish messages; i.e., `get_rotation` is now a **callback** function. A callback function is an "infinite loop" that publishes its results to subscribers at specific intervals. Line 17 links the */odom Odometry* topic to the `get_rotation` callback function. Any object subscribing to the `Odometry` topic will receive the latest `get_rotation` message. Line 19 specifies that the messages will be published every second. Lines 20–23 are our "infinite loop," which returns the current orientation every second in quaternions [x, y, z, w].

Executing the rotateRobotOdom.py

Execute the following Linux terminal commands to compile and run in terminal 1:

```
$ cd ~/catkin_ws/
$ catkin_make
$ source devel/setup.sh
$ cd ~/sim_gazebo_rviz_ws/
$ catkin_make
$ source devel/setup.sh
$ cd ~/sim_gazebo_rviz_ws/
$ roslaunch ai_rover_worlds ai_rover_cat.launch
```

After success, execute the following Linux terminal commands in terminal 2:

```
$ cd ~/catkin_ws/
$ rosrun rotate_robot rotateRobotOdom.py
```

After executing ai_rover_cat.launch and rotateRobotOdom.py, you will see the AI rover in the Gazebo with a scrolling list of radians in the terminal. We see that the initial direction that the AI rover faces is approximately zero radians. Figure 7-1 shows our initial heading, and all future headings will be based upon this original orientation.

Figure 7-1. *The initial orientation angle of around 0 radians for AI rover*

If we want to control the rover with the keyboard, we must execute the teleops_twist_keyboard.py ROS node. We can do this with the following terminal commands in terminal 3:

```
$ cd ~/sim_gazebo_rviz_ws/src
$ roslaunch ai_rover_remastered_description
  ai_rover_teleop.launch
```

Using the "j" and "l" keys, turn the rover and take note of the values displayed in terminal 2. These numbers reflect the orientation of the rover in radians.

Revisiting our **exploratory** mission, let us imagine that our rover is moving forward when the sensors detect an object. The sensing of the object is the feedback. Continuing forward would be wrong because we would crash, so we have to change the rover's direction—issue new commands. To avoid the object, we can turn left into an unblocked, unexplored area; turn right into an unblocked, unexplored location; or turn around and retreat to an earlier point with unexplored regions. This last option recognizes that we hit a dead end and that further exploration is futile.

It is crucial to verify that the rover's orientation on the screen is correct. Zero radians should look like the rover is pointing "North" (up) on the screen. From now on, if we command the AI rover to turn (yaw), we should see the radians change (Figure 7-2), and the rover's orientation will change on the screen. In the background, our command is converted into independent commands for each wheel. In response, after processing the command, each wheel publishes its current position and yaw via a wheel encoder. Each encoder type is inaccurate, but it is accurate enough for our system. So, our rover's pose is calculated from the wheel positions.

In Figure 7-2, we changed the orientation of the rover to ~0.16 radians with the manual controls (j, k, l). We can see that we have now successfully tested, verified, and validated that the odometry and orientation of the AI rover are indeed correct.

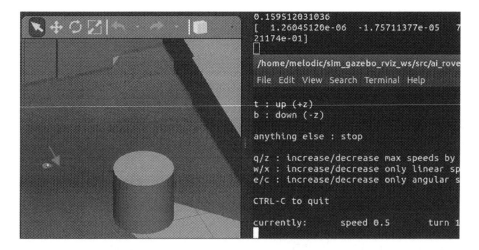

Figure 7-2. *Changes of the yaw orientation in radians. Orientation is 0.16 radians and is facing south–southeast (SSE)*

Control Theory

Control theory is simply a feedback loop (Figure 7-3); i.e., we do something, then look around to see how things have changed because of what we did. If the result is good, we continue doing it. If the result is wrong, we adjust and "correct" what we are doing wrong. For example, we start our rover one-point exploratory mission by moving forward. As long as the sensors detect nothing in front, we continue to move forward. However, once any of our sensors (LiDAR, radar, camera) detect an obstacle in our path, we change the AI rover's course to avoid the obstacle and explore further.

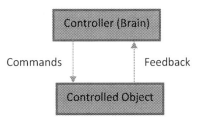

Figure 7-3. *Simple feedback loop*

Autonomous Navigation

Figure 7-4 uses a simple Unified Modeling Language (UML) statechart to model navigation and control logic for our one-point exploration mission; i.e., we have no goal location or resource. We are just trying to create a map. The UML statechart models the three states (GO_STRAIGHT, STOP, and TURN). A limited number of state transitions (obstacle, no obstacle) will represent the navigation's dynamic behavior. If the rover is going straight and there are no obstacles in its path, it will keep going straight. However, if a sensor detects an obstacle, it will stop and then turn. While it is turning, if there is still an obstacle in its path, it will keep turning. When there is a clear (and unexplored) path, it will resume going forward down that unexplored path. Not shown in this diagram is what happens when no new unexplored paths are available. One option is to maintain a set of visited waypoints, retrace the path back to a previous point with unexplored paths, and resume exploring from that point.

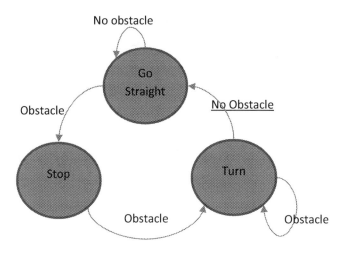

Figure 7-4. *Navigation statechart*

From a programming perspective, the three states must update odometry data. Later, we will also use the states to update our internal map. We will review these concepts for autonomous navigation toward the end of this chapter. At the end of this chapter, we will review the following behaviors that the rover can accomplish when exploring the environment.

> **GO_STRAIGHT:** The rover starts or continues going in its current direction, continuously updating its position <x,y,z> [z never changes]. Roll, pitch, and yaw <r,p,y> are not changed. If it detects an obstacle, it stops.

> **STOP:** Upon entering the STOP state, we add this location <x,y,z> to the visited waypoint list. (If we have external location references, this is the time to correct odometry location errors.) We now transition to the TURN state.

TURN: The rover changes its heading to "look" in a direction not explored, updating the yaw <r,p,y> [roll and pitch do not change], but the position <x,y,z> stays the same. While turning, we look for unobstructed, unexplored paths. When we find an unobstructed, unexplored path, we store the direction as exploring and return to the GO_STRAIGHT state. If there are no unexplored paths, we backtrack to a previous waypoint and continue exploring from there.

We will leave to you what modifications are necessary for uneven terrain; i.e., changes to <z> and <r,p>.

In our statechart, we chose to ignore tracking of waypoints visited. As we will soon see, each state in the statechart directly maps to functions in our SLAM results in the navigational stack and the move_base process.

Simultaneous Localization and Mapping (SLAM)

We use SLAM for two reasons: to generate a map and to locate the rover (and objects) in the generated map. Mapping occurs during an exploration mission. The rover will have to sense and avoid detected objects during the map generation, such as walls, columns, etc. After generating the map, the rover can traverse "optimal paths" that avoid all objects. The second reason to use SLAM means locating the rover relative to other objects in the environment. The final generated map will have a "global" coordinate system that the rover's "local" coordinate system can overlay.

SLAM maps can be dynamic and evolve, but we will initially assume the environment is static and flat (no holes) so as to understand the SLAM algorithm better. When dynamic, moving objects are introduced, the map will need to change over time. The map will have to categorize objects as permanent or transitory. Our understanding means that the generated map is "final" and complete.

SLAM's exploratory algorithm (exploratory mission), called Adaptive Monte Carlo Library (AMCL), allows our AI rover to explore its environment. A Monte Carlo method randomly chooses a direction to turn toward, like rolling dice or spinning a roulette wheel. If the selected random direction becomes obstructed, it "rolls again." The SLAM algorithm creates the initial map by randomly wandering around the environment. As it explores, it fills in the map with a tentative solution with objects (walls, stationary objects). It "gains confidence" in the map's structure as it continues its exploration. Initially, the generated map will not detect holes in the floor, such as stairs. Since the GoPiGo has a simple proximity sensor, we can repurpose it to detect holes in front of the rover. We will add that sensor script later.

One of the cool things about SLAM is generating paths to backtrack based on the *partially* generated map. SLAM tracks "decision points"—i.e., locations the rover turned—as the rover explores. If the rover arrives at a dead end—i.e., all directions from that decision-point location have been explored—SLAM will generate a path back to a previous decision-point location to begin exploring again. If that last location is now fully explored, it will backtrack further, and so on. This implies that SLAM will track how completely explored each decision-point location is. Once all decision-point locations are thoroughly investigated, the mapping process for the environment is completed. This completed exploration does not mean that the rover touched every spot in the environment.

Now we have enough information to perform goal-oriented missions. Given the rover's initial location and the goal point, SLAM can generate intermediate waypoints to define the path the rover should traverse. Assuming the explored area is a flat environment, the rover follows the path.

Installing SLAM and Associated Libraries

Installing SLAM libraries will look complicated, but it is not. Essentially, we install SLAM libraries and libraries that use SLAM. The essential library is gMapping; most other SLAM libraries call upon it. There are many types of SLAM libraries. We are using OpenSLAM, and you can read about it at `https://OpenSLAM.org`.

OpenSLAM's `slam_gmapping` library supplies the required ROS interface wrapper between the sensors, ROS, and the gMapping software. The `slam_gmapping` provides a 2D grid map (similar to developing the floorplan on grid paper). The gMapping library has two data sources: odometry and LiDAR. The odometry data provides localization (context) to the gMapping library; i.e., the rover's current location based on its starting point. The LiDAR gives object-detection data relative to the current rover location. Thus, `slam_gmapping` transforms each LiDAR scan into the odometry transform (TF) frame of the AI rover. Therefore, as the AI rover moves and explores, we see each LiDAR scan update the map details.

We must install the Noetic ROS openslam's `gMapping` package and related support packages. The support packages include sensor, exploration, image processing, and transformation ROS packages. To do this, we must enter the following commands in a Linux terminal:

```
$ sudo apt-get upgrade
$ sudo apt-get install ros-noetic-laser-proc ros-noetic-rgbd-
launch ros-noetic-depthimage-to-laserscan
$ sudo apt-get install ros-noetic-rosserial-arduino ros-noetic-
rosserial-python ros-noetic-rosserial-server ros-noetic-
rosserial-client ros-noetic-rosserial-msgs
$ sudo apt-get install ros-noetic-compressed-image-transport
ros-noetic-rqt-image-view
$ sudo apt-get install ros-noetic-gmapping ros-noetic-ros-
noetic-interactive-markers
```

```
$ sudo apt-get install ros-noetic-turtle-tf2 ros-noetic-tf2-
tools ros-noetic-tf
$ sudo apt-get install ros-noetic-slam-gmapping
$ sudo apt-get install ros-noetic-hector-slam
$ sudo apt-get install ros-noetic-rtabmap-ros
$ sudo apt-get install ros-noetic-teleop-twist-keyboard
$ sudo apt-get install ros-noetic-amcl
$ sudo apt-get install ros-noetic-move-base
$ sudo apt-get install ros-noetic-map-server
```

Setting Up SLAM

We'll be using the stable Noetic ROS version of SLAM. To process the examples of Noetic ROS SLAM, we will need to make the following changes in the system.

Setting Up the Noetic ROS Environment

Please navigate to your Linux home directory and open the .bashrc file by running the following terminal command in Terminator:

```
$ cd
$ nano .bashrc
```

Please add the following elements at the end of the .bashrc file and source the file to allow SLAM to work correctly:

```
export ROS_MASTER_URI=http://localhost:11311/

export ROS_HOSTNAME=localhost
$ source /opt/ros/noetic/setup.bash
```

We have upgraded Noetic ROS and its supporting dependencies to run SLAM and its associated libraries and facilities.

Initializing the Project Workspace

```
H$ source /opt/ros/noetic/setup.bash

$ cd ~/catkin_ws
$ source devel/setup.bash
$ catkin_make
```

Note   Sourcing ROS and a ROS SLAM project directory is critical
to running the project. Running `roscore` in the background is
extremely important.

Navigational Goals and Tasks

Although the current rover can sense and avoid objects, it can easily get
trapped. To avoid this, the rover must navigate using course correction to
map the maze entirely. For our purposes, we will move forward until an
object obstructs our path. We will then either (1) turn to avoid the object in
a direction we have never explored or (2) retrace our "steps" to a location
where there are directions we have not explored.

Create a Python ROS node script to read odometry data (yaw and
position) from each simulated wheel encoder. This information includes
the estimated precision and accuracy values. Therefore, we will now
develop the heading correction control required to allow the AI rover to
travel from one location point to the next destination point. This heading
correction will be the first actual navigational task for the AI rover. This
navigational task of going from one waypoint to the next will only be a
linear navigational track and correction program. This type of navigational
track means that the AI rover will not avoid and circumvent obstacles
at this stage of development, while going from one location point to the

next location point. We will develop this linear navigational tracking and correction first and then create the sense-and-avoid capabilities later in the chapter to develop ***and test*** each component for the AI rover in an incremental and modular fashion. We will also reveal and review each wheel encoder's simulated data location within this ROS node program. Finally, there will also be a brief discussion in this section on using simple mathematical objects such as quaternions. These objects will allow us to develop more reliable and efficient algorithms to determine the correct heading and tracking for the AI rover.

We must update the folder directories in order for our first navigational tracking and correction algorithm to be able to process and send information to and from multiple sources. These input sources include a simulated PixHawk autopilot or encoders in the form of an odometry source topic (`'/ai_rover_remastered/base_controller/odom/'`). The navigation trajectory commands then go to the (`'/ai_rover_remastered/base_controller/cmd_vel'`) output topic for the AI rover's trajectory control.

Importance of Maps

Why does our rover require a map in the first place? A fully developed map represents the locations of important objects in the environment. A map gives context. Not all objects will necessarily be represented, just the important ones. However, some maps can still be wrong because objects on a map are not detected. For example, if you were to look at the map generated by our SLAM of the rover world, you would see four dots forming a rectangle. These are not four objects evenly spaced in a rectangle, but instead the dumpster wheels. SLAM can still create incorrect map representations. That is why a camera is needed to determine what objects are in the pathway of the rover.

Initially, the rover knows nothing about its environment. It must map the environment. Unfortunately, a "fog of war" (FoW) covers the entire map, in military parlance. The objective of the exploratory mission is to reduce the FoW, ideally to zero percent.

The rover randomly walks around and maps the environment on its first mission. The rover does NOT touch every location in the environment, but rather, it "sees" walls and obstacles in the distance and marks them on the map. It does not walk up to the wall. After this mission is complete, we have a map with the locations of barriers and obstacles. SLAM can now generate optimal paths from an initial position to a goal position, with waypoints created to traverse the now "known" environment safely.

SLAM gMapping Introduction

The algorithm serving as the foundation for our SLAM gMapping is the Rao-Blackwellized particle filter. SLAM uses this particle filter to outline the map's boundaries based on the LiDAR sensor's laser-range data and the rover's odometry. This filter uses a probabilistic approach that is beyond this book's scope to explain.

The `slam_gmapping` ROS node subscribes to the `/odom` and `sensor_msgs/LaserScan` topics and publishes the occupancy grid and map to `nav_msgs/OccupancyGrid`. The occupancy grid is the size of the map supplied by the user. We will use a 40x40 occupancy grid ranging between -20 and +20. Think of the occupancy grid as a map written onto a piece of paper. The drawing of the map must remain on the grid.

The `slam_gmapping` node combines the LiDAR and odometry local data and transforms it into a global map overlayed onto the occupancy grid. The `remap` line in the following launch file connects `slam_gmapping` to the rover, thus connecting the LiDAR and odometry sensors. From the LiDAR sensor's point of view, it does not move; it is mounted at a fixed location on the rover. The rover moves, but the sensor does not "know" this! It just does its job of scanning. This movement of the LiDAR means the laser data must be translated from its local fixed location (on the moving rover) to its actual global position on the map. Each laser scan is offset by the rover's current position, calculated using odometry data.

Recall that the rover does not know what the environment looks like at the beginning of its journey. Its initial view is a "blank" map that needs exploration; i.e., fog of war. The `slam_gmapping` starts to fill in the map as the rover moves around the environment. To do this, `slam_gmapping` subscribes to the `sensor_msgs/LaserScan` and `/Odom` (`tf/tfMessage`) topics published by the rover's LiDAR and Odom sensors. The tf message topic changes the senor data from local rover coordinates to global map coordinates.

Launching Our Rover

We have two launch files to run our rover using SLAM: `ai_rover_world.launch` and `gmapping_demo.launch`. Additionally, we will need a world map called `rover.world`. Since designing a world map is beyond the book's scope, download it from the book website.

Creating ai_rover_world.launch

To begin the mapping, first launch our rover (Listing 7-2) in its Gazebo environment (source file follows):

```
$ roscore
$ roslaunch ai_rover_remastered ai_rover_world.launch
```

Figure 7-5 displays the rover inside the Gazebo environment, indicated by the small dot inside the red box. Notice how the size of the environment dwarfs our rover. We expect this small scale of size.

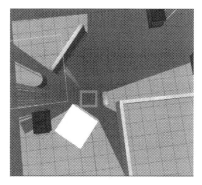

Figure 7-5. *Rover now spawned (in the red box)*

Listing 7-2. The ai_rover_world.launch File

```xml
<?xml version="1.0" encoding="UTF-8"?>
<launch>
 <arg name="world" default="empty"/>
 <arg name="paused" default="false"/>
 <arg name="use_sim_time" default="true"/>
 <arg name="gui" default="true"/>
 <arg name="headless" default="false"/>
 <arg name="debug" default="false"/>

 <include file="$(find gazebo_ros)/launch/empty_world.launch">
  <arg name="world_name" value="$(find ai_rover_remastered)/
  worlds/rover.world"/>
  <arg name="paused" value="$(arg paused)"/>
  <arg name="use_sim_time" value="$(arg use_sim_time)"/>
  <arg name="gui" value="$(arg gui)"/>
  <arg name="headless" value="$(arg headless)"/>
  <arg name="debug" value="$(arg debug)"/>
 </include>
```

```
<param name="robot_description" command="$(find xacro)/xacro
'$(find ai_rover_remastered)/urdf/ai_rover_remastered.xacro'"/>

<node name="ai_rover_remastered_spawn" pkg="gazebo_ros"
type="spawn_model" output="screen" args="-urdf -param robot_
description -model ai_rover_remastered" />

</launch>
```

Now, launch the slam_gmapping map builder in a separate terminal. The ros.org gmapping_demo.launch was modified to more closely align with our environment and rover. The source code is in the **"Modified gmapping_demo.launch File"** section.

```
$ roslaunch ai_rover_navigate gmapping_demo.launch
```

The slam_gmapping Launch File

The slam_gmapping ROS node processes sensor data and generates a map while the rover explores its environment. The gMapping file has several parameters that we can set in the launch file. Most of these we do not modify; underlined parameters have changed:

- **base_frame (default: "base_link"):** This is the frame's name attached to the rover's mobile base.

- **map_frame (default: "map"):** This is the frame's name attached to the map. This is also the name of the topic we use in RViz.

- **odom_frame (default: "odom"):** The frame's name of the odometry system. We set the odometry system for the rover's physical differential-wheel-drive encoder or Gazebo-simulated plug-in drive.

- **map_update_interval (default: 5.0):** Wait time (in seconds) until the next map update. This number is important since a shorter wait-time duration could cause system degradation of the rover or simulation.

- **maxRange (float):** Sets the maximum range of the laser. Set this value to the real-world range of the LiDAR.

- **maxUrange (default: 80.0):** Sets the maximum usable range of the laser. The laser beams will be stopped at this range of distance.

- **minimumScore (default: 0.0):** Sets the minimum score (distance from an object) to have an accurate laser reading.

- <u>**xmin**</u> **(default: -100.0, set: -20):** The minimum x-range of the map. Set xmin as close as possible to the actual total x-range of the environment that the rover will explore, which in our case is 40 meters. So we will set xmin to -20 and xmax to 20 meters.

- <u>**ymin**</u> **(default: -100.0, set: -20):** The minimum y-range of the map.

- <u>**xmax**</u> **(default: 100.0, , set: 20):** The maximum x-range of the map.

- <u>**ymax**</u> **(default: 100.0, set: 20):** The maximum y-range of the map.

- <u>**delta**</u> **(default: 0.05, set: -0.01):** Resolution of the map.

- <u>**linearUpdate**</u> **(default: 1.0, set: 0.5):** The required linear distance in the (x-direction) that the rover must move to process laser readings.

- **angularUpdate (default: 0.5, set: 0.436):** The required angular distance the rover must move to process laser readings.

- **temporalUpdate (default: -1.0):** Wait time (seconds) between laser readings. If this value is -1.0, it turns off this function, i.e., continuous.

- **particles (default: 30, set: 80):** Number of filter particles.

- **resampleThreshold (default: xx, set: 0.5):** The sensor data frequency in seconds.

Note   We need to ensure that we shut down the previously launched slam_gmapping node by pressing CTRL+C on the same terminal before every simulation.

Preparing slam_gmapping Package

We must organize our SLAM processing software into ROS packages. We will now create an ROS ai_rover_navigation package. This particular package will contain a mixture of code (e.g., ROS nodes), data, libraries, images, documentation, and so forth. Every SLAM program will be included in an ROS package. An ROS package aims to provide proper functionality and encourage the reuse of gmapping_demo.launch for other ROS systems in the rover. Use the following terminal commands to create the ai_rover_navigation package:

```
$ cd ~/catkin_ws/src
$ catkin_create_pkg ai_rover_navigation std_msgs rospy roscpp
```

```
$ cd ~/catkin_ws/src/ai_rover_navigation
$ mkdir launch
$ cd ~/catkin_ws/src/ai_rover_navigation/launch
```

The package has three dependencies: std_msgs, roscpp, and rospy. The std_msgs are common data types that have been predefined in ROS. In ROS 1, the ROS libraries are two independent libraries written in C++ (roscpp) and Python (rospy). Both libraries do NOT have the same functionality and are not equivalent! The SLAM and ROS libraries depend on both of these libraries.

Modify gmapping_demo.launch File

Download the gmapping_demo.launch from the ros.org website.

```
$ gedit gmapping_demo.launch
```

Modify the file from the ROS website to match our rover's parameters (Listing 7-3).

Listing 7-3. The gmapping_demo.launch File

```
<?xml version="1.0"?>
<launch>
 <master auto="start"/>
 <param name="/use_sim_time" value="true"/>

 <!--- Run gmapping -->
 <node pkg="gmapping" name="slam_gmapping" type="slam_gmapping"
output="screen">

   <!--- Occupancy Grid --->
   <param name="delta" value="0.01"/>
   <param name="xmin" value="-20"/>
   <param name="xmax" value="20"/>
```

```
<param name="ymin" value="-20"/>
<param name="ymax" value="20"/>

<param name="linearUpdate" value="0.5"/>
<param name="angularUpdate" value="0.436"/>
<param name="temporalUpdate" value="-1.0"/>
<param name="resampleThreshold" value="0.5"/>
<param name="particles" value="80"/>

<!--- Connect to Rover's Lidar -->
<remap from="scan" to="ai_rover_remastered/LaserScan/Scan "/>
<param name="base_frame" value="base_link" />

</node></launch>
```

The last section of the launch file connects the gMapping package to the rover-transformed LiDAR data; i.e., the data is now in global coordinates. This launch file does not display anything on the screen. To do that, we need to use RViz (from the rover's perspective) and Gazebo (from the operator's perspective). (We will launch this file two sections from now.)

RViz gMapping

To display the generated map in RViz, we need a launch file (ai_rover_rviz_gmapping.launch) to connect the rover sensors, RViz, and the map (Listing 7-4).

Listing 7-4. The ai_rover_rviz_gmapping.launch File

```
<?xml version="1.0"?>
<launch>

<param name="robot_description" command="$(find xacro)/
xacro '$(find ai_rover_remastered_description)/urdf/ai_rover_
remastered.xacro'"/>
```

```
<!-- send fake joint values -->
<node name="joint_state_publisher" pkg="joint_state_publisher"
type="joint_state_publisher">
 <param name="use_gui" value="False"/>
</node>

<!-- Combine joint values -->
<node name="robot_state_publisher" pkg="robot_state_publisher"
type="robot_state_publisher"/>

<!-- Show in Rviz   -->
<node name="rviz" pkg="rviz" type="rviz" args="-d $(find ai_
rover_description)/rviz/mapping.rviz"/>
<!--node name="rviz" pkg="rviz" type="rviz" args="-d $(find
ai_rover _description)/launch/ai_rover.rviz"/-->

</launch>
```

Since the code for defining the RViz mapping.rviz parameters is so large, it can be found in this textbook's GitHub source code repository. The parameter descriptions for mapping.rviz can also be found in the RViz configuration file.

To control the rover, use the keyboard Teleop script. Remember to click on the appropriate terminator screen running the Teleops program to move the robot. As you move your rover, you should see that the SLAM gMapping process is updating the map in RViz with new features of the unexplored regions. Use the keyboard keys shown in Figure 7-6 to control the rover in the environment.

To begin the process of manually controlling the rover, use the Teleops program with the following shell command:

```
rosrun teleop_twist_keyboard teleop_twist_keyboard.py
```

You should see something like Figure 7-7. The slam_gmapping ROS node updates the map with an "object" (blue box). The red lines are potential "walls," and the orange box is the intrepid rover.

i	Move forward
,	Move backward
j	Turn left
l	Turn right
k	Stop
q z	Increase / Decrease Speed

Figure 7-6. *The basic keyboard commands to move the rover*

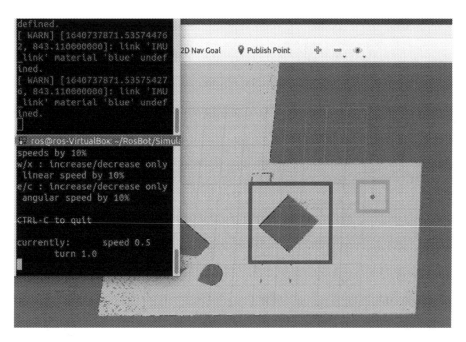

Figure 7-7. *The slam_gmapping node updates the map as the rover explores*

After fully exploring the environment, we save the generated map image (pgm) and metadata (yaml). We do this by entering the following command in a terminator shell window:

```
rosrun map_server map_saver -f ~/ai_rover_remastered/maps/
test_map
```

Convert the test_map.pgm into a JPG in the same directory. You can now view it in an image viewer, such as gimp (Figure 7-8).

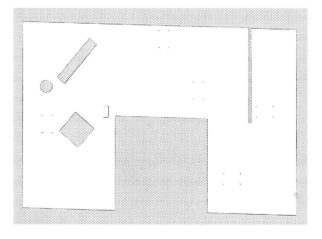

Figure 7-8.   *The final gMapped image file*

The pgm file is a "picture" of the map, while the yaml file describes the size of the map. The yaml file description comes later.

Final Launch Terminal Commands

To begin the mapping, first launch the rover in its environment. We do that with the following commands in different terminals. In the first terminal, we launch rover into the environment (existing hidden map):

```
$ roscore
```

```
$ roslaunch ai_rover_remastered ai_rover_world.launch
```

Next, we add the gMapping package that enables the rover to explore the environment and create a map by launching the slam_gmapping map builder in a second terminal:

```
roslaunch ai_rover_navigate gmapping_demo.launch
```

Finally, in a third terminal, we open an RViz window:

```
roslaunch ai_rover_remastered ai_rover_rviz_gmapping.launch
```

Your RViz display should be similar to Figure 7-9

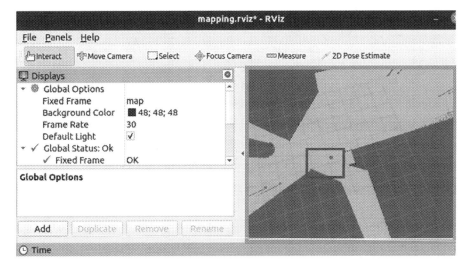

Figure 7-9. *Initial RViz in Gazebo with laser scan data (red points). The red line in the blue box appears to be a wall, but this is an error*

RViz Mapping Configurations

We now turn our attention to mapping for ROS navigation. We should also focus on the critical components needed for successful ROS mapping and navigation of environments. We need to review what is necessary to configure RViz to display map and navigation information and data from the rover. We will also discuss what is required to configure RViz to enable our intrepid rover's successful mapping and navigation. We will also review the many possibilities of creating a map with our rover. Why is mapping so critical for the navigation of our rover? Because mapping will allow our rover to plan pathways and avoid collisions with objects in our unexplored environment.

Our rover can only access maps from two sources during its exploration mission. The first source is the map already provided to the rover by the mission planner. The second source is for the rover to construct its very own map from sensor (LiDAR) and odometry data. Creating a map from the rover's sensor (LiDAR) and odometry data is called the SLAM (Simultaneous Location and Mapping) process. One very critical tool needed to monitor our intrepid rover is RViz. RViz is the ROS graphical environment responsible for monitoring information, messages, and data sent between the rover and human operator. The importance of RViz as a visual data environment is especially true for mapping.

Now we will review the steps necessary to visualize the `LaserScan` data and the map of the environment. To display the mapping data visually on RViz will require us to obtain three sources or topics from the rover. We need to get the `LaserScan` display, `Odometry` data display, and the `Map` display as options in RViz. We will also need to add the robot description model developed first in URDF source files in Chapter 4 and further enhanced with object-oriented features with Xacro extensions in Chapter 5. We will need to display the robot description model of the rover to see the `LaserScan` results so as to visually detect any abnormalities.

We first need to execute the `gmapping_demo.launch` file, which launches the `slam_gmapping` ROS node. We need to have this ROS node running first in order to obtain the messages and data sources necessary for the map. The following are the terminal commands required for starting the mapping process:

```
$ cd ~/{primary user-defined rover directory}
$ source devel/setup.bash
$ catkin_make
```

First Terminator Shell Window Enter (From Chapter 5):

```
$ roslaunch ai_rover_remastered ai_rover_world.launch
```

Second Terminator Shell Window Enter:

```
$ roslaunch ai_rover_remastered_navigation gmapping_demo.launch
```

Third Terminator Shell Window Enter:

```
$ rosrun rviz rviz
```

Note   If you have an issue with the rosrun rviz rviz command, you'll need to import our intrepid rover's required robot description model using the RViz launch file ai_rover_remastered_rviz.launch from Chapter 5. There, you can import the correct robot description model of our rover. Start RViz with roslaunch ai_rover_remastered ai_rover_remastered_rviz.launch.

If you successfully start the ai_rover_remastered_rviz.launch file from Chapter 5, then you will see the following figure displayed in RViz (Figure 7-10).

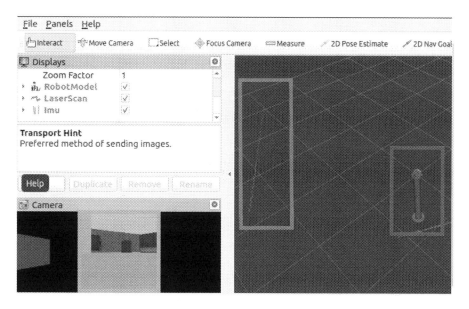

Figure 7-10. *Successful start of the RViz environment with our intrepid rover and sensor data display boxes (camera–orange, LaserScan–red, and IMU–blue)*

Checking LaserScan Configurations

First, we must examine the LaserScan display to inspect any displayed errors. Please search for similar solutions on ROS.org and other supporting websites if errors are detected. Examine the Displays options in RViz and choose the LaserScan one (Figure 7-11). You should see that we have the correct LaserScan topic of /ai_rover_remastered/laser_scan/scan. The rover's RobotModel display is included. We modify the LaserScan size in meters and so forth. For more information regarding the LaserScan topic, go to: http://wiki.ros.org/laser_pipeline/Tutorials/IntroductionToWorkingWithLaserScannerData.

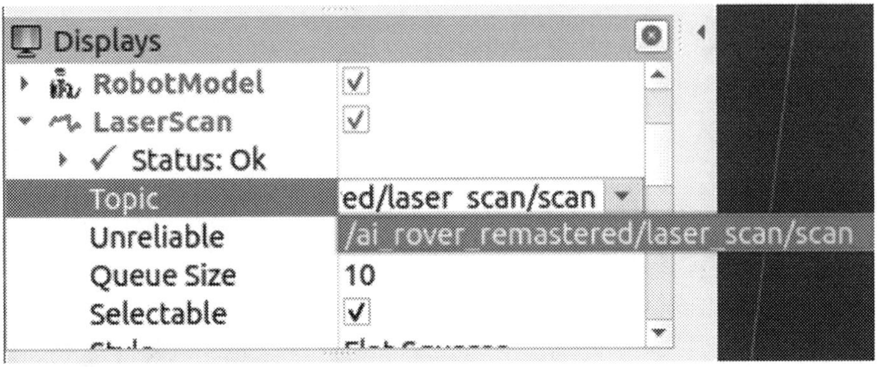

Figure 7-11. *LaserScan RViz option with the correct topic of /ai_rover_remastered/laser_scan/scan. Topic defined in ai_rover_remastered.xacro file*

Checking Mapping Configurations

If we have our rover correctly imported into RViz, we should also see the Gazebo world that the rover is exploring. To do this, we must go to the Display options in RViz and select the Add button located to the bottom left of the Display options. Click on the Add button and add the Map display option. We then go to the Map display properties and set our topic to /map. We will now visualize the gray-colored map generated by the slam_gmapping ROS node, the raw image from the RGB camera onboard our rover, and the IMU data in RViz (Figure 7-12).

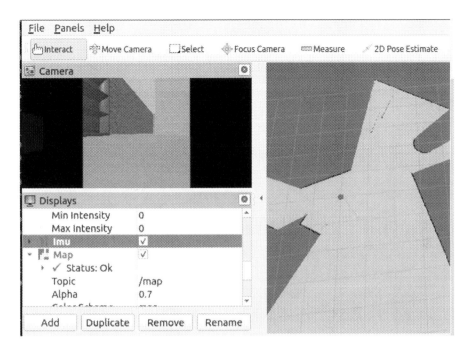

Figure 7-12. *Camera, LaserScan, IMU, and now gray-colored slam_gmapping ROS node mapping information displayed in RViz. Our rover is becoming ever more sophisticated and capable of autonomous navigation*

Now let us examine another RViz feature by changing some of the elements of the various RViz displays to visualize different aspects of the Gazebo simulation. We can modify the angle of the LaserScan, number of scans, etc. All of these changes can affect the generated mapping. Please try different values for the attributes of the camera, LaserScan, and IMU topic displays as an exercise. We have a fully working slam_gmapping ROS node displaying mapping data in RViz. We must now save our RViz configuration for future experiments.

Saving RViz Configurations

RViz also can save concurrent robot description and sensor topic configurations very quickly. This capability of RViz allows us to quickly recover our configurations at a later time. The following are the steps that enable us to save our RViz configurations:

- First, go to the top-left corner of the RViz environment screen and select the File menu.

- Second, select "Save Config As" with the left mouse button. You will also need to save the configuration file as default_cam_lidar_IUM.rviz on the desktop. Once you save this file to the desktop, you can then move this file to the /rviz folder under the ai_rover_remastered ROS package. See steps one and two in Figure 7-13.

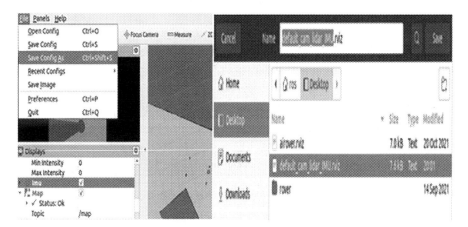

Figure 7-13. *The left display has the "Save Config As" option highlighted. The right display has the default_cam_lidar_IMU.rviz saved on the desktop*

We now have the default_cam_lidar_IMU.rviz file saved and moved to the correct /rviz folder directory under the leading ai_rover_remastered ROS package directory. We should test to see if we can start RViz with the correct configurations already enabled. To test if the RViz configurations are indeed saved correctly, please exit out of all currently running terminals by typing CTRL+C for each terminal. Once the terminals have stopped, we can proceed and re-enter the following terminal commands:

```
$ cd ~/{primary user-defined rover directory}
$ source devel/setup.bash
$ catkin_make
```

First Terminator Shell Window Enter (From Chapter 5):
```
$ roslaunch ai_rover_remastered ai_rover_world.launch
```
Second Terminator Shell Window Enter:
```
$ roslaunch ai_rover_remastered_navigation gmapping_demo.launch
```
Third Terminator Shell Window Enter:
```
$ rosrun rviz rviz
```

Once we have RViz running, we can then go to it and reopen our saved RViz configuration file. We can do this with the following steps:

- First, go to the top-left corner of the RViz environment screen and select the File menu.

- Second, select and click "**Open Config**" with the left mouse button. Then, open default_cam_lidar_IUM.rviz on the desktop. You will then have your configuration file opened with all sensors, topics, and the mapping generation included with RViz.

Note   If you again have problems with the `rosrun rviz rviz` command, you once again need to import our intrepid rover's required robot description model using the RViz launch file `ai_rover_remastered_rviz.launch` from Chapter 5. There, you import the correct robot description model of our rover. Start RViz with `roslaunch ai_rover_remastered ai_rover_remastered_rviz.launch`. We will also need to reset our map by adding it as a topic and selecting the topic's name to /map. This action is documented in Figure 7-14.

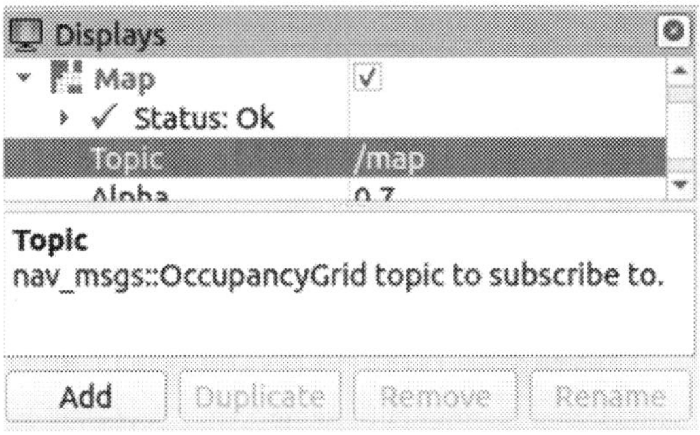

Figure 7-14. *We are resetting our Map display (if required) with the topic /map*

Ok. Now that we have our RViz correctly configured, we need to execute our `teleops_twist_keyboard` command by opening another shell in the Terminator shell program and once again entering the following command:

```
$ rosrun teleop_twist_keyboard teleop_twist_keyboard.py
```

Once you start this program, you should be able to manually control the rover to explore the environment with your keyboard. There may be some areas you are not able to explore. Do not worry about this issue. Explore the places you can map. Also, please notice the generated map by comparing the features developed in the map and the Gazebo environment. You want to ensure no "ghosts," anomalies, or missing regions on the map. Missing areas on the map could prevent the rover from exploring the entire environment.

Additional Noetic SLAM Information

Before we review additional mapping features with Noetic ROS, we should remember two essential items, SLAM and `slam_gmapping` ROS node. SLAM is the algorithm responsible for building a map of an environment while also keeping track of where our rover is located in that same environment. The SLAM algorithm solves the "chicken-or-the-egg" issue of mapping and localization of the rover. ROS Noetic uses the gMapping algorithm as a `slam_gmapping` ROS node to spare the robotics developer from developing the same algorithm again. The encapsulation of `slam_gmapping` as an object in our SLAM allows us to use LiDAR and odometry pose data provided by our rover moving in the environment. The `slam_gmapping` ROS node subscribes to the `LaserScan` and odometry topics, transforms the rover's dimensions, and creates the occupancy grid map (OGM). An occupancy grid map is either a 2D or a 3D array of cells, each cell of which stores a number. In the case of our intrepid rover, we will only use a 2D array of number cells. The number in each cell indicates the probabilistic likelihood that a cell contains an obstacle. The number ranges from 0 (free space) to 100 (100 percent occupied). The LiDAR's unscanned areas are marked as a -1. These are the items created when we launched the `gmapping_demo.launch` file. However, some potential mapping issues and failures with `gmapping_demo.launch` will occur if any of the five following conditions occurs:

1. A mapping error will occur if the physical rover's LiDAR system falls off the rover or is no longer at the rover's center during operation. If this happens, please place the LiDAR back in its original place specified in the rover's URDF or Xacro specification description files.

2. A mapping error will occur if the rover model in Gazebo and RViz is not the same as the physical rover (GoPiGo3). Be sure that the simulated rover's dimensions and moment of inertia closely approximate the actual physical rover. Failure to do so will cause the rover's tf, LiDAR, and odometry information to not be reproducible in the real-world rover.

3. We will need to make multiple changes to the rover's URDF or Xacro files and plug-ins **reflecting any changes** to the rover's features or sensors, such as LiDAR. We will need to update any changed or affected sensors or topics used in RViz, for example.

4. A mapping error will occur if we suddenly replace the primary LiDAR sensor with a different sensor, such as a radar or a binocular camera, and make no changes to the underlying URDF, Xacro, or plug-in files.

5. A mapping error could happen if we use another physical LiDAR sensor system with different characteristics than the Hokuyo LiDAR used in our Gazebo simulation. Differences between the physical and simulated LiDAR could be scan rate, angle sweep, frequency parameters, etc. All of these subtle differences between LiDARs could lead to mapping anomalies.

The map_server ROS Node

Another critical component contained inside of ROS is the map_server node. This ROS node offers the map data as an ROS service. This node also allows the saving of generated maps to a file. This ROS node also provides the map data to any requesting ROS node. For example, an ROS node that handles navigation could request the most recent available map. The move_base ROS node finalizes the request to get the most recent map data for pathfinding or localizing our rover in an environment. The following is an ROS service that provides the map occupancy grid data for our generated map: static_map (nav_msgs/GetMap).

Apart from requesting the map through the preceding service, there are two latched, or the last saved, topics that you can connect to get an ROS message with the map. These saved topics can provide the previously saved message, even if there is no more map data. The topics with which this node writes the map data are as follows:

- map (nav_msgs/OccupancyGrid): This topic provides the map occupancy data.

- map_metadata (nav_msgs/MapMetaData): Provides the map metadata

Map Image Savings or Modifications

We will review how to save and modify the maps that ROS Noetic SLAM generates. We will use the map_server package to save, process, and modify maps generated by the slam_gmapping ROS node. The map_server ROS package also includes the map_saver ROS node, allowing us to save and modify the map data using an ROS service. The map_saver ROS package saves the current map and occupancy grid and creates two files. The first file created is the map.pgm file, which has map data, including the occupancy grid data (free space, obstacle[s], and unknown). The second file created is the `map.yaml` file, which includes the metadata of the occupancy grid and the image name of the map. We have already gone through the process of saving a map. However, we will now detail these two essential files generated by the `map_server` ROS package.

Again, we should `source devel/setup.bash`, `catkin_make` compile, and launch the `ai_rover_remastered_gazebo.launch`, `gmapping_demo.launch`, `ai_rover_remastered_rviz.launch`, and the `teleops_keyboard_twist` script files, each in its very own terminal shell, as before, in Terminator. These individual commands and printout listings can be seen in Figure 7-15.

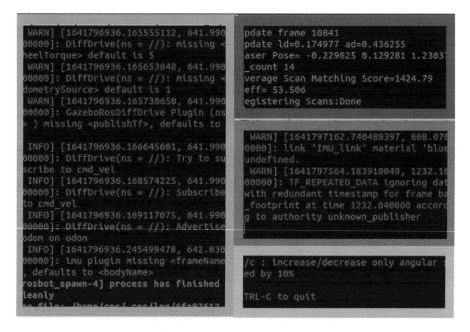

Figure 7-15. *The terminal runs ai_rover_remastered_gazebo.launch (orange), gmapping_demo.launch (blue), ai_rover_remastered_rviz. launch (red), and teleops_keyboard_twist (green)*

By doing this, we can save the map by using an additional separate terminal under the /src root directory and entering the following shell command:

```
$rosrun map_server map_saver -f rover_map
```

This command will create rover_map.pgm and rover_map.yaml files under the main /src root directory.

rover_map.pgm Map Image File Data

We need to inspect the rover_map.pgm file generated in the previous section. We must do the following steps:

1) Save the rover_map.pgm file under the main /src directory.

2) If you do not have a gimp image editor, please install it. Use the gimp image editor to open the image stored in the main /src directory. The terminal command needed to open the file is the following:

```
$ sudo apt-get install gimp
$ gimp rover_map.pgm
```

3) Now you can visualize, modify, and save the image if required. The rover_map.pgm can be seen in Figure 7-16.

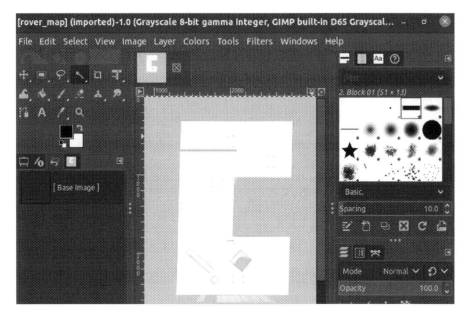

Figure 7-16. *The Gimp image editor with the rover_map.pgm file*

The rover_map image describes the occupancy grid of the entire world in the (white, black, or dark gray) color of the corresponding pixels. Color and grayscale images are compatible, but most maps are white, black, and dark gray (even though these PGM images may be stored in color). White pixels are free-space, black pixels are obstacles, and any dark-grey pixels in between are unscanned areas.

When communicated via ROS topic messages, each occupancy grid element is represented as a range of numbers from 0 to 255 (8-bit), where 0, being free-space, to 255 is wholly occupied.

rover_map.yaml Map File Metadata

To view the generated rover_map.yaml file from an earlier section, go to the main workspace /src folder and see whether the rover_map.yaml file is present. You will need to use a simple text editor to view this file.

You can view the information inside of the rover_map.yaml file by typing the following command: $vi rover_map.yaml. The data contents of the rover_map.yaml file can be seen in Figure 7-17.

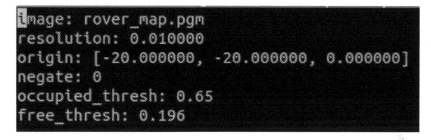

```
image: rover_map.pgm
resolution: 0.010000
origin: [-20.000000, -20.000000, 0.000000]
negate: 0
occupied_thresh: 0.65
free_thresh: 0.196
```

Figure 7-17. *The generated rover_map.yaml file*

Also, when you start the following terminal command—$ rosrun map_server map_server rover_map.yaml—you will process and receive the map metadata and be able to examine ROS topics, as in Figure 7-18.

```
$ rosrun map_server map_server rover_map.yaml
[ INFO] [1641959957.589132701]: Loading map from image "
rover_map.pgm"
[ INFO] [1641959957.702176948]: Read a 4000 X 4000 map @
 0.010 m/cell
```

Figure 7-18. *Metadata of ROS topics of rover_map.yaml*

We then have the following contained inside of rover_map.yaml:

- **Image:** Name of the file (rover_map.pgm) containing the image of the generated map.

- **Resolution:** Resolution of the map (in meters/pixel).

- **Origin:** Coordinates of the lower-left pixel in the map. These coordinates are given in 2D (x,y). So our coordinates are set to -20.0 for x and -20.0 for y. The third value indicates rotation. The value zero means no rotation.

- **Occupied Thresh:** Pixels with a value greater than this (0.65 or 65% probability of an object's being present) will be considered obstacles.

- **Free Thresh:** Pixels with a value smaller than this (0.196 or 19.6% probability of an object's being present) will be considered free space.

- **Negate:** Inverts the colors of the map. By default, white means completely free and black means dangerous or obstacle present.

ROS Bags

We now can see that Noetic ROS also allows us to create a map in real-time as the rover explores the environment. This process is why we move the rover (slowly and avoiding quick turns) as data processing occurs while mapping the domain. Making maps is based on the published LiDAR and transformed (odometry) topics. If we need to extract published data on these topics, we will need to use a bag file to publish that extracted data into a map. A bag is a file format in ROS for storing ROS message data during a rover mission. For more information regarding ROS bags, please refer to the following link: `http://wiki.ros.org/Bags`. There are two steps for generating maps of bag-file data.

We first take action to create the actual bag file for the published topic data. Before continuing, close all terminal processes in ALL terminals with the CTRL+C command. We then need to launch the `ai_rover_remastered` Gazebo simulation and the `teleops_keyboard_twist` program in separate terminals. So in terminal one, we have the following commands:

```
$ cd ~/catkin_ws
$ source devel/setup.bash
$ catkin_make
$ cd ~/catkin_ws/ai_rover_remastered
```

```
$ roslaunch ai_rover_remastered_description ai_rover_
remastered_gazebo.launch
```

We then launch terminal two, where we have the following commands:

```
$ cd ~/catkin_ws
$ source devel/setup.bash
$ catkin_make
$ rosrun teleop_twist_keyboard teleop_twist_keyboard.py
```

We then launch terminal three, where we have the following commands:

```
$ cd ~/catkin_ws
$ source devel/setup.bash
$ catkin_make
$ rostopic list
```

We should now be controlling the rover in the environment, and we should be careful not to make quick turns so as to allow the published data to be collected correctly in the bag file. Please ensure the rover overlaps its start and end points during the entire mapping process. We should get the following list of topics by entering the previous rostopic command (not a complete rostopic list):

/ai_rover_remastered/laser_scan/scan
```
/clock
/cmd_vel
/gazebo/link_states
/gazebo/model_states
/gazebo/parameter_descriptions
/gazebo/parameter_updates
/gazebo/performance_metrics
/gazebo/set_link_state
/gazebo/set_model_state
/imu
```

```
/odom
/rosout
/rosout_agg
```
/tf

We need to ensure that our intrepid rover still publishes valid data for bold-faced **LaserScan** and **tf** topics. Now that we are controlling our rover with the Teleops program, we need to start recording the **LaserScan** and **tf/transforms** data into the bag file we created earlier.

Once we have the bag file, we can build our map. We need to start the slam_gmapping node, which will process laser scans for the topic /ai_rover_remastered/laser_scan/scan. We do this with the command:

```
$ rosbag record -O roverlaserdata /ai_rover_remastered/laser_scan/scan /tf
```

In a separate terminal, we run the following command:

```
$ rosrun gmapping slam_gmapping scan:= ai_rover_remastered/laser_scan/scan
```

We then open another terminal and enter the following:

```
$ rosbag play roverlaserdata
```

We will need to use the map_server to create the necessary map files (.pgm and .yaml) to generate the map. Now we will enter the following command to create the map:

```
$ rosrun map_server map_saver -f rovermap
```

Importance of ROS Bags

The ROS "rosbag" toolset has the purpose of recording, storing, and replaying data that our rover sends and receives. The data may include LiDAR, radar, camera images, and teleop_twist_keyboard commands.

The following are examples of multiple applications that use this archived sensor data:

- Archived sensor data can recreate the environment in the simulation. The data would help for debugging purposes.

- Archived sensor data aids machine learning, such as training neural networks for perception and cognition.

- Archived data can be used to determine the causality of disastrous events for the autonomous rover.

Localization (Finding Lost Rovers)

Let us now find the location of our intrepid rover. To perform correct and safe navigation with our rover, we need to know the rover's location and orientation. Its orientation would be the direction that it is facing. We need our map to navigate autonomously with the rover, but it is worthless without knowing our rover's location and direction in the context of the SLAM-generated map. Let's look at a quick demonstration of localization. The most widely used algorithm for localization is the Adaptive Monte Carlo Localization (AMCL) algorithm. Eventually, we will need to build script programs based on AMCL to allow the rover to locate itself on a map.

Adaptive Monte Carlo Localization (AMCL)

The Adaptive Monte Carlo Localization (AMCL) algorithm, which serves as the foundation for SLAM, is based on the particle filter. We will keep our description brief, but essentially a particle filter sends out many particles or samples spanning an entire search space. The number of particles helps to reduce uncertainty with the layout and boundaries produced by SLAM. The search space is our rover's environment it is trying to map and explore. For more information regarding the particle filter or the Adaptive Monte Carlo Localization algorithm, please refer to the following references:

```
https://en.wikipedia.org/wiki/Particle_filter
https://roboticsknowledgebase.com/wiki/state-estimation/
adaptive-monte-carlo-localization/
```

Now, how does the AMCL algorithm work for the localization of our rover? To use AMCL for localization, we start with a map of our environment. This map is the same map generated in Figure 7-22. We can either set the rover to some known position, in which case we are manually localizing it, or make the robot start from no initial estimate of its position. By manually localizing the rover, we place the rover in an exact position where the laser scan overlaps the boundary outlines of our map. As the robot moves forward, we generate additional readings that estimate the robot's pose after the motion command. Sensor readings are incorporated by reweighting these samples and normalizing the weights. Generally, adding a few random uniformly distributed samples is good as it helps the robot recover when it has lost track of its position. In those cases, the robot will resample from an incorrect distribution without these random samples and never recover. It takes the filter multiple sensor readings to converge because, within a map, we might have dis-ambiguities due to symmetry in the map, which gives us a multi-modal posterior belief (Figure 7-19).

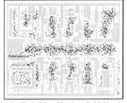
Fig. 2: Global localization: Initialization.

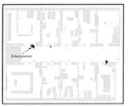
Fig. 3: Ambiguity due to symmetry.

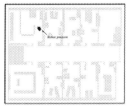
Fig. 4: Successful localization.

Figure 7-19. *Distribution of random samples*

Localizing the rover is essential for navigating from one pose (position and orientation) to another on the map. Now that we can localize our rover in the map, we can configure the localization system for our intrepid rover and determine its current pose. We will now check all possible parameters used for the amcl_demo.launch: all SLAM gMapping launch, YAML, and script files can be found in the GitHub account for this publication. The AMCL node is also highly reconfigurable, and we can easily define each parameter with different values. Also, in addition to setting these parameters in the amcl_demo.launch file, we can set these parameters in a separate YAML file referenced by the amcl_demo.launch file.

We will now define the general, filter, and laser parameters for the amcl_demo.launch file:

- **odom_model_type** (default: "diff"): Uses the odometry model of the rover. It can be "diff," "omni," "diff-corrected," or "omni-corrected" as possible options.

- **odom_frame_id** (default: "odom"): The odometry frame.

- **base_frame_id** (default: "base_link"): The rover base frame.

- **global_frame_id** (default: "map"): Indicates the name of the coordinate frame published by the localization system.

- **use_map_topic** (default: false): Indicates whether the node gets the map data from the topic or a service call.

- **min_particles** (default: 100): Sets the filter's minimum allowed number of particles.

- **max_particles** (default: 5000): Sets the maximum allowed number of particles for the filter.

- **kld_err** (default: 0.01): Set the maximum error between the true and estimated distribution.

- **update_min_d** (default: 0.2): Sets the linear distance (in meters) that the robot has to move to perform a filter update.

- **update_min_a** (default: $\pi/6.0$): Sets the angular distance (in radians) that the robot has to move to perform a filter update.

- **resample_interval** (default: 2): Sets the number of filter updates required before resampling.

- **transform_tolerance** (default: 0.1): Time (in seconds) to postdate the published transform to indicate that this transform is valid into the future.

- **gui_publish_rate** (default: -1.0): Maximum rate (in Hz) at which scans and paths are published for visualization. If this value is -1.0, this function is disabled.

- **laser_min_range** (default: -1.0): Minimum scan range to be considered; -1.0 will cause the laser's reported minimum range to be used.

- **laser_max_range** (default: -1.0): Maximum scan range to be considered; -1.0 will cause the laser's reported maximum range to be used.

- **laser_max_beams** (default: 30): How many evenly spaced beams in each scan are used when updating the filter.

- **laser_z_hit** (default: 0.95): Mixture weight for the z_hit part of the model.

- **laser_z_short** (default: 0.1): Mixture weight for the z_short part of the model.

- **laser_z_max** (default: 0.05): Mixture weight for the z_max part of the model.

- **laser_z_rand** (default: 0.05): Mixture weight for the z_rand part of the model.

Configuring the AMCL ROS Node

At the conceptual level, the AMCL package maintains a probability distribution over the set of all possible robot poses, and updates this distribution using data from odometry and laser range finders.

The AMCL package represents the probability distribution using a particle filter at the implementation level. The filter is "adaptive" because it dynamically adjusts the number of particles in the filter: when the rover's pose is highly uncertain, it increases the number of particles; when the rover's pose is well determined, it decreases the total number of particles. This "cloud of green arrows" can be seen by the size of the green arrows around the rover in RViz becoming larger or smaller as the rover's uncertainty increases or decreases. In Figure 7-20, there will be two RViz windows, the environment map, the rover on that map, and many green arrows. Those green arrows represent estimations of the rover on the map. The green arrows are estimations that the localization algorithm is doing to figure out the rover's location on the map. The green arrows will concentrate on the most likely rover site when moving the robot. The package also requires a predefined map of the environment against which to compare observed sensor values. This comparison enables the robot to make a trade-off between processing speed and localization accuracy.

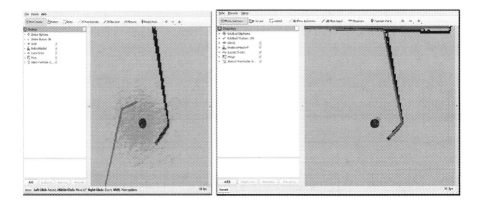

Figure 7-20. *On the left, we can see the laser scan of rover (red), the large "cloud" of green arrows, and the outline of a map boundary (black). On the right, we can see that once we have a correct pose for the rover, with our laser scan overlapping a map boundary, our uncertainty "cloud" of green arrows is minimized*

You can assist the rover in being located on the map more efficiently by providing it its location on the map. To do that, go to the RViz map window. Then, press the 2D Pose Estimate button, go to the map, and point at the rover's approximate location. The AMCL package can autonomously determine the rover's position without manual rover placement, but manually placing the rover with its approximate pose (position and orientation) is a more efficient use of time and allows the AMCL package to converge quicker.

Even though the AMCL package works fine out of the box, there are various parameters that one can optimize based on one's knowledge of the platform and sensors. Configuring these parameters can increase the performance and accuracy of the AMCL package and decrease the recovery rotations that the rover carries out while navigating.

Three ROS parameters can configure the AMCL node: overall filter, laser model, and odometry model. These three parameters are edited and located in our demo amcl.launch file.

Here is a sample **nonworking** launch file. We will be developing the
amcl_demo.launch file (Listing 7-5) in the next source code listing. The
following launch file is merely an example of the parameters used by
the AMCL algorithm. Generally, you can leave many parameters at their
default values.

Listing 7-5. Nonworking amcl_demo.launch File

```
<?xml version="1.0"?>
<launch>`
    <arg name="map_file" default="$(find
    ai_rover_navigation)/maps/rover_map.yaml"/>
    <node name="map_server" pkg="map_server" type="map_server"
    args="$(arg map_file)" />
    <arg name="use_map_topic" default="true"/>
    <arg name="scan_topic" default="/ai_rover_remastered/
    laser_scan/scan" />
    <node pkg="amcl" type="amcl" name="amcl">
    <param name="use_map_topic" value="$(arg use_map_topic)"/>
     <!-- Publish scans from best pose at a max of 10 Hz -->

    <param name="odom_model_type" value="diff"/>
    <param name="odom_alpha5" value="0.1"/>
    <param name="gui_publish_rate" value="10.0"/>
    <param name="laser_max_beams" value="60"/>
    <param name="laser_max_range" value="12.0"/>
    <param name="min_particles" value="500"/>
    <param name="max_particles" value="2000"/>
    <param name="kld_err" value="0.05"/>
    <param name="kld_z" value="0.99"/>
    <param name="odom_alpha1" value="0.2"/>
    <param name="odom_alpha2" value="0.2"/>
```

```
<!-- translation std dev, m -->
<param name="odom_alpha3" value="0.2"/>
<param name="odom_alpha4" value="0.2"/>
<param name="laser_z_hit" value="0.5"/>
<param name="laser_z_short" value="0.05"/>
<param name="laser_z_max" value="0.05"/>
<param name="laser_z_rand" value="0.5"/>
<param name="laser_sigma_hit" value="0.2"/>
<param name="laser_lambda_short" value="0.1"/>
<param name="laser_model_type" value="likelihood_field"/>
<!-- <param name="laser_model_type" value="beam"/> -->
<param name="laser_likelihood_max_dist" value="2.0"/>
<param name="update_min_d" value="0.25"/>
<param name="update_min_a" value="0.2"/>
<param name="odom_frame_id" value="odom"/>
<param name="resample_interval" value="1"/>
<!-- Increase tolerance because the computer can get quite
busy -->
<param name="transform_tolerance" value="1.0"/>
<param name="recovery_alpha_slow" value="0.0"/>
<param name="recovery_alpha_fast" value="0.0"/>
<remap from="scan" to="$(arg scan_topic)"/>
```

```
</node>
</launch>
```

The purpose of this AMCL_demo.launch is to show a nonworking launch file for the AMCL node with all of the possible parameters listed. We will now develop an actual **working** amcl_demo.launch file (Listing 7-6) that uses some of the parameters listed in the previous amcl_demo.launch file. Please note that this operating version of amcl_demo.launch is also launching a second node. This second ROS node is the move_base node.

This node is responsible for both local and global cost maps for the rover. These cost maps will allow us to have navigation and obstacle-avoidance capabilities. We will review and apply the abilities of the navigation stack, move_base, and cost maps for navigation in the **"Programming Goal Locations for Rover"** section. The **working** amcl_demo.launch file for localizing the rover uses the following script.

Listing 7-6. Working amcl_demo.launch File

```xml
<?xml version="1.0"?>
<launch>
  <master auto="start"/>

  <!-- Map server -->
  <arg name="map_file" default="$(find ai_rover_remastered_
  navigation)/maps/rover_map.yaml"/>
  <node name="map_server" pkg="map_server" type="map_server"
  args="$(arg map_file)" />

  <!-- Place map frame at odometry frame -->
  <node pkg="tf" type="static_transform_publisher" name="map_
  odom_broadcaster"
      args="0 0 0 0 0 0 map odom 100"/>

  <!-- Localization -->
  <node pkg="amcl" type="amcl" name="amcl" output="screen">
    <remap from="scan" to="/ai_rover_remastered/laser_
    scan/scan"/>
    <param name="odom_frame_id" value="odom"/>
    <param name="odom_model_type" value="diff-corrected"/>
    <param name="base_frame_id" value="base_link"/>
    <param name="update_min_d" value="0.5"/>
    <param name="update_min_a" value="1.0"/>
  </node>
```

```xml
<!--include file="$(find amcl)/examples/amcl_omni.launch"/-->

<!-- Move base -->
<node pkg="move_base" type="move_base" respawn="false"
name="move_base" output="screen">
  <rosparam file="$(find ai_rover_remastered_navigation)/
  config/costmap_common_params.yaml" command="load"
  ns="global_costmap" />
  <rosparam file="$(find ai_rover_remastered_navigation)/
  config/costmap_common_params.yaml" command="load"
  ns="local_costmap" />
  <rosparam file="$(find ai_rover_remastered_navigation)/
  config/local_costmap_params.yaml" command="load" />
  <rosparam file="$(find ai_rover_remastered_navigation)/
  config/global_costmap_params.yaml" command="load" />
  <rosparam file="$(find ai_rover_remastered_navigation)/
  config/base_local_planner_params.yaml" command="load" />

  <remap from="cmd_vel" to="cmd_vel"/>
  <remap from="odom" to="odom"/>
  <remap from="scan" to="/ai_rover_remastered/laser_
  scan/scan"/>
  <param name="move_base/DWAPlannerROS/yaw_goal_tolerance"
  value="1.0"/>
  <param name="move_base/DWAPlannerROS/xy_goal_tolerance"
  value="1.0"/>

</node>
</launch>
```

Importance of Localization and AMCL

The rover's localization in an environment is of utmost importance. However, we need to know what localization and navigation mean in ROS. For example, we need to know the internal mechanisms that allow our ROS to localize our rover in the environment, which also serves as the foundation for ROS navigation. The rover's localization problem is especially critical as the rover moves around the environment map. ROS needs to know the rover's position and orientation, for without knowing these attributes, navigation is impossible. ROS localization of the rover also depends on the rover's continuous stream of sensor readings. These readings allow us to continually reduce the uncertainty of the rover's pose (position and orientation) as the rover moves with time. We will see how RViz and the AMCL node reduce the rover's pose uncertainty.

Visualizing the AMCL in RViz

The first action item in ROS is to make certain that all processes in all terminals have been shut down or killed with the CTRL+C command before starting another simulation. This action is critical to prevent anomalies from occurring with ROS. Once we have halted any running terminals, we can begin the rover Gazebo simulation with the following commands on the first opened terminal in Terminator:

```
$ cd ~/catkin_ws
$ source devel/setup.bash
$ catkin_make
$ cd ~/catkin_ws/ai_rover_remastered
$ roslaunch ai_rover_remastered_description ai_rover_
remastered_gazebo.launch
```

We need to execute the actual AMCL launch file in a separate running terminal to launch the **AMCL** node. We need to have this node running to visualize the pose arrays. We start the AMCL ROS node with the following terminal command:

```
$ cd ~/catkin_ws
$ source devel/setup.bash
$ catkin_make
$ cd ~/catkin_ws/ai_rover_remastered
$ roslaunch ai_rover_remastered_navigation amcl_demo.launch
```

Now that we have launched the amcl_demo.launch script file, we can develop the RViz launch file that will launch the RViz environment with all of the correct connections necessary for displaying results from the amcl_demo.launch program. We develop this launch file with the following commands in a third terminal in Terminator:

```
$ cd ~/catkin_ws
$ source devel/setup.bash
$ catkin_make
$ cd ~/catkin_ws/ai_rover_remastered
$ cd launch
$ gedit ai_rover_remastered_amcl_rviz.launch
```

We have now developed the ai_rover_remastered_amcl_rviz. launch file. This file is similar to the RViz launch file from Chapter 5. It allows us to set waypoints in RViz for the rover to follow autonomously. Also, we will see that the RViz node has a configuration file called $(find ai_rover_remastered)/rviz/amcl.rviz. The amcl.rviz file sets all of the configurations necessary for RViz to handle information from the amcl_demo.launch file. However, this configuration file is substantially large and will not be listed to save chapter space. We will, however, review specific limited sections of the amcl.rviz file that are important for you to comprehend. The amcl.rviz file will once again be available on the supporting website for this book. We can now enter the following source code for the ai_rover_remastered_amcl_rviz.launch file (Listing 7-7).

Listing 7-7. The ai_rover_remastered_amcl_rviz.launch File

```
<?xml version="1.0"?>
<launch>
    <param name="robot_description" command="$(find xacro)/
    xacro$(find ai_rover_remastered)/urdf/ ai_rover_
    remastered.xacro"/>

    <!-- send fake joint values -->
    <node name="joint_state_publisher" pkg="joint_state_
    publisher" type="joint_state_publisher">
    <param name="use_gui" value="False"/>
    </node>
    <!-- Combine joint values -->
    <node name="robot_state_publisher" pkg="robot_state_
    publisher" type="robot_state_publisher"/>

    <!-- Show in Rviz   -->
    <node name="rviz" pkg="rviz" type="rviz" args="-d $(find
    ai_rover_remastered)/rviz/amcl.rviz"/>
</launch>
```

We now need to launch the RViz environment in a separate fourth terminal. We will need to modify and add the following necessary displays to the RViz environment:

- **Map Display**, as shown in the previous sections

- **LaserScan Display** from the LiDAR sensor

- **PoseArray Display** for the AMCL analysis of the map

317

We can now start the RViz environment with the `ai_rover_`
`remastered_amcl_rviz.launch` file. Please activate RViz with the following
terminal commands in a separate fifth running terminal:

```
$ cd ~/catkin_ws
$ source devel/setup.bash
$ catkin_make
$ cd ~/catkin_ws/ai_rover_remastered_navigation
$ roslaunch ai_rover_remastered_navigation ai_rover_remastered_
amcl_rviz.launch
```

Once the RViz environment activates, we must add the **PoseArray**
display with the following actions in RViz:

- Click the Add button (blue box) under Displays and
  select the PoseArray display under the Topics tab
  (orange box). This action is shown in Figure 7-21.

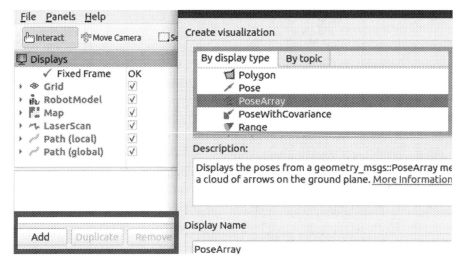

Figure 7-21. *The PoseArray display for RViz*

Now that we have the RViz GUI interface open, we should have the rover Gazebo simulation and the `amcl_demo.launch` program running, and this RViz GUI available to you. If all three of these programs are running, we should see the display found in Figure 7-22. We need to add the `/particlecloud` topic for the PoseArray display type. Once we add this topic to the PoseArray display type, we should see a "cloud" of red arrows around the blue box found inside of Figure 7-22. We should see this cloud of red arrows get smaller as our uncertainty decreases. The more the rover moves, the less uncertain the rover's pose will become. Our rover's location is at the center of this very same blue box. The orange box in Figure 7-22 is where our previously generated map's features overlap the laser scan of the rover. The green box is where our PoseArray display type is visible (Figure 7-22).

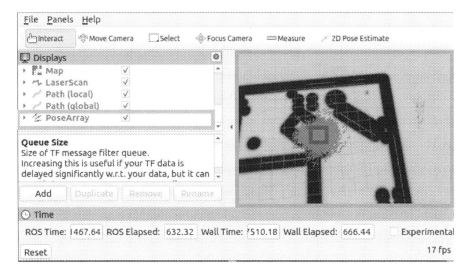

Figure 7-22. *The rover's red arrow/particlecloud overlaps the blue box*

Now we need to test that our rover's pose is correct. We will need to create an ROS service server that will determine the rover's current pose (position and orientation) at that exact moment. We will need to create a

ROS package currentPose that will include rospy, roscpp, and std_msgs as a package dependency. The following will create our ROS package by opening another terminal in Terminator:

```
$ cd ~/catkin_ws
$ source devel/setup.bash
$ catkin_make
$ cd ~/catkin_ws/src
$ catkin_create_pkg currentPose std_msgs rospy roscpp
$ cd ~/catkin_ws/src/currentPose
$ mkdir launch scripts
$ cd ~/catkin_ws/src/ai_rover_navigation/scripts
$ gedit findPose.py
$ chmod +rwx findPose.py
```

Now enter the source code for findPose.py (Listing 7-8):

Listing 7-8. The findPose.py File

```python
#! /usr/bin/env python3

import rospy
from std_srvs.srv import Empty, EmptyResponse
from Empty.srv.
from geometry_msgs.msg import PoseWithCovarianceStamped, Pose

robot_pose = Pose()

def service_callback(request):
    print("Rover Pose:")
    print(rover_pose)
    return EmptyResponse() # the service Response class, in
    this case EmptyResponse

def sub_callback(msg):
```

```
    global robot_pose
    robot_pose = msg.pose.pose

rospy.init_node('service_server')
my_service = rospy.Service('/get_pose_service', Empty ,
service_callback) # create the Service called get_pose_service
with the defined callback
sub_pose = rospy.Subscriber('/amcl_pose',
PoseWithCovarianceStamped, sub_callback)
rospy.spin() # mantain the service open.
```

We will then need to start the launch file for determining the current pose of the rover. The following is the list of terminal commands for a new terminal window in Terminator:

```
$ cd ~/catkin_ws
$ source devel/setup.bash
$ catkin_make
$ cd ~/catkin_ws/src
$ cd ~/catkin_ws/src/currentPose/launch
$ gedit currentPose.launch
$ chmod +rwx currentPose.launch
```

Then we enter the source code launch file (Listing 7-9).

Listing 7-9. The currentPose.launch File

```
<launch>
    <node pkg="currentPose" type="findPose.py" name="service_
    server" output="screen">
    </node>
</launch>
```

Moving the Rover's Pose with RViz

We have the rover in our Gazebo environment, and we have the PoseArray fully activated. We can move the rover from one pose (position and orientation) to another. But first, we need to make certain the rover is in the correct initial pose. We look to the actual pose of our rover in the Gazebo environment and try to closely match that same pose for the same rover in RViz. We make corrections to the rover by using the **2D pose Estimate** button located at the top of the RViz GUI window. Once the corrections happen, we can select the **2D Nav Goal** option and click on a reasonably close position to the rover. Also, make certain that the goal pose you choose is in free space (free of any obstacle). Once that happens, you should see a dotted blue line from the initial pose to the goal pose. This blue line is a global path, which is the optimal distance between the initial rover pose and the final goal pose. The line is the shortest path possible between these two points. Once the rover begins to travel to the goal pose, we will see a green line trail behind the rover, the local path. We can see this in Figure 7- 23.

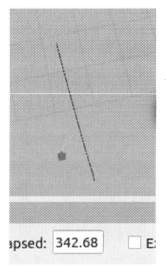

Figure 7-23. *Blue line is global path, and faint green line is local path*

Programming Goal Poses for Rover

We are far from finished allowing our rover to perform ROS navigation. For now, we need to start navigating autonomously in ROS. The first step toward ROS navigation was to create a map. The second step allowed the rover to accurately localize and determine its pose (position and orientation) in the environment. Now we need to command where and how the rover needs to go. Part of telling the rover where to go on the map is developing the rover's correct path planning. Path planning takes the current pose of the rover and the goal pose where the rover needs to go and attempts the shortest pathway to reach that goal pose as an output.

Now we have a localized rover in the simulated Gazebo environment. We have directly sent it goal poses (positions and orientations) through RViz. We have happily seen the rover traverse with these goal poses across the environment. Ultimately, this navigation by goal-pose waypoints is possible by interacting with the ROS navigation system. This navigation system is also referred to as the navigation stack. For more information regarding the navigation stack, please refer to `http://wiki.ros.org/navigation/Tutorials`. Now we will quickly review the functions and characteristics of the navigation stack and what makes it operate. These functionalities will allow us to use Python scripts to control the rover without direct human interaction in RViz.

Noetic ROS Navigation Stack

We are far from finished allowing our rover to perform ROS navigation. For now, we need to start navigating autonomously in ROS. The first step toward ROS navigation was to create a map. The second step allowed the rover to determine its current pose in the environment accurately. The third step was to send goal-pose waypoints to the rover in RViz. Now we will discuss the purpose of the navigation stack.

The navigation stack is the most heavily used component of ROS. It is a central component to our rover that allows the rover to move about the world to different waypoints without colliding with objects. The navigation stack integrates the map, localization system, sensors (LiDAR), and odometry to plan from the initial pose to the final pose. Also, the navigation stack can allow the rover to recover from issues of becoming trapped in the environment by rotating around the rover's Z-axis until it recovers.

A high-level description of the navigation stack is as follows:

- First, send a navigational goal to the navigation stack. Then, make an ROS action call with a goal type MoveBaseGoal, which specifies the final goal pose, usually in the map coordinate frame.

- Second, the navigation stack calculates the shortest possible pathway from the initial pose to the goal pose using the map's pathfinding algorithm in the global planner.

- Third, the navigation stack then passes the global pathway solution to the local planner. The local planner then attempts this pathway solution, using the rover's sensors to avoid obstacles that might not be present on the map. If the local planner fails, the global planner can issue a new pathway solution.

- Fourth, as the rover approaches the goal pose within a given distance, the rover arrives at its destination and the action terminates.

Configuring the Navigation Stack

To develop a Python script that controls the rover's navigation with a known map we need to launch three important ROS nodes:

- **The move_base node** handles the global pathfinding and local control of the rover.

- **The amcl node** localizes the rover concerning the referenced map.

- **The map_server node** provides the static map for the rover to localize and plan.

Summary

The objective of this chapter was to develop the necessary skills to create ROS navigational capabilities for our intrepid rover. These skills include generating and storing maps of the environment (catacombs), localizing the rover in our simulated environment, allowing the rover's Python programs to perform path planning, establishing navigational waypoints visualizing data, and correcting any simulation errors in RViz. This chapter has also cited numerous references that can enhance understanding of ROS navigation. It has also used multiple ROS navigation packages. ROS navigation concepts were used to allow us to develop our rover's capabilities further with the following features:

- Utilize a basic navigation structure by using the already available and standard ROS navigation stack.

- Create and store a map of the environment generated by the `slam_gmapping` ROS node.

- Utilize the `move_base` ROS node to set waypoints.

- Localize our rover in the environment.

- Develop a pathfinding solution in the environment.

- Travel the path solution and avoid obstacles along the way.

CHAPTER 8

OpenCV and Perception

Any autonomous rover will need to know what objects lie in its path while exploring the Egyptian catacombs. While the LiDAR can help identify objects, it can only "see" objects at the level of its LiDAR sweep. The rover misses detecting any object lower than the LiDAR sweep. It also misses any object that hangs from the ceiling that does not bisect the LiDAR sweep. We need a more robust system, called computer vision. Unfortunately, it is computationally expensive. Computer vision mimics the way humans detect objects. Our rover needs the ability to extract information from images and recognize objects by their patterns and features. The rover must process pixels and colors to determine edges, helping it traverse the environment and avoid obstacles.

Objectives

The following are the objectives required for successful completion of this chapter:

- Understand computer vision basics

- Install OpenCV and the necessary connections to Robotic Operating System (ROS).

© David Allen Blubaugh, Steven D. Harbour, Benjamin Sears, Michael J. Findler 2022
D. A. Blubaugh et al., *Intelligent Autonomous Drones with Cognitive Deep Learning*,
https://doi.org/10.1007/978-1-4842-6803-2_8

- Use color filtering for vision processing

- Use edge detection to help find walls and objects in the catacombs

- A brief introduction to convolutions and their relations with computer vision and convolutional neural networks.

- Use of morphological transformations for image and object processing

Overview

In the previous chapter, we supplied the rover with limited situational awareness. However, there are several situations in which the rover might crash because the LiDAR did not detect an object; for example, short objects. Adding computer vision to the rover enhances its understanding of its immediate environment.

The previous chapters focused on the LiDAR as the only sensor on the rover. We used the LiDAR system as the foundation for sense-and-avoidance routines in Chapter 6, and we used the LiDAR platform to map and navigate an environment. We will now use OpenCV image processing in coordination with the LiDAR to improve the sense-and-avoid algorithm. Furthermore, we will develop a feedback-loop system to determine misalignments and correct spatial errors. All of the examples in previous chapters had no error-correcting, so the location of the rover was an approximation. This approximation got worse the longer the rover was moving on a mission.

Recall that we installed the camera driver in Chapter 5; now, we will use it. We will be using the OpenCV library. ROS considers the OpenCV library as an ROS node.

Introduction to Computer Vision

Computer vision is the scientific field that deals with the interaction of computers and how they receive, process, and interpret digital images or video. From the perspective of the rover, the use of computer vision will allow the rover to understand the environment through visual processing. The computer vision applications for this rover include acquiring, processing, and understanding digital images from either a sensor (camera) in a Gazebo simulation or an RGB camera in the real world. Computer vision is also a critical component in machine learning (deep learning) applications for producing numerical information used in subsequent decision-making algorithms (deep reinforcement learning). Computer vision analysis uses geometry, physics, statistics, and machine learning theory models to process and make decisions regarding acquired digital imagery.

The computer vision for the rover seeks to apply models of digital imagery to make decisions based on images. Fundamentally speaking, computer vision is implemented as an electronic system. This electronic system is composed of an input sensor link, such as a camera; a central processing system to process the input information; and an output device, such as an actuator, that receives output commands from the main processor. The central processing system usually makes decisions based on the digital imagery taken from the input camera and then sends output commands to the actuator. Digital imagery takes many forms, such as video sequences, views from multiple cameras, or multi-dimensional data from a 3D scanner. The sub-domains of computer vision that our intrepid rover will use are the following:

- Object detection

- Use of color filters

- Use of edge detectors

- Motion estimation

- Distance estimation

The following sections introduce the three related fields, which form the foundations of the computer vision analysis used in the interaction of ROS and OpenCV for this chapter.

Solid-State Physics

Solid-state physics is the knowledge used to allow solid-state materials to construct sensors, such as RGB electro-optical cameras and LiDAR. Solid-state physics is closely related to computer vision. Most computer vision systems rely on image sensors, which detect electromagnetic radiation, typically in either the visible or the infrared spectrum.

Neurobiology

Neurobiology is a field of applied research science that is the basis for learning-based methods developed for computer vision. This field of science in neural medicine has created the sub-field within computer vision where artificial systems are designed to emulate the behavior and structure of biological vision systems (typically human eyes) at different levels of complexity. This biology structure includes the interconnected network structures of neural networks often seen in human vision.

Robotic Navigation

Robot navigation sometimes involves autonomous path planning or deliberation so that robotic systems can navigate an environment. A detailed understanding of these environments is required to navigate through them. A computer vision system provides critical information about the rover's immediate environment. The rover has now become a mobile vision sensor platform.

What Is Computer Vision?

OpenCV (open source computer vision) is a programming library aimed at real-time computer vision. Intel originally developed OpenCV, but Willow Garage now maintains it. OpenCV supports GPU acceleration for real-time operations on the Raspberry Pi 4 and the Nvidia Jetson embedded systems.

In the previous chapter, the rover extracted data from the environment. Unfortunately, the data from the LiDAR was sparse and one pixel "wide," which could cause problems (getting stuck under a dumpster or stalactite). To flesh out the data stream, we need to add another source. Thus, we introduce the camera. It has two purposes: to be a rich source of data for the operator and to provide the images for the OpenCV library. The first usage is raw data projected to an operator's console for human input to the rover. The second usage will be the initial input to OpenCV and generate 2D edges of objects in the environment. This essentially takes the one-dimensional data of the LiDAR (horizontal) and adds a second dimension (vertical). The image processing is slower than the LiDAR processing, so that it will be called less frequently (10–20 times per second compared to 30–60 times per second for the LiDAR).

The rover cannot identify objects that rise from the floor (stalagmites) or sink from the ceiling (stalactites) and do not intercept the LiDAR sensor. These short objects may still intersect with the rover chassis, so they must be avoided. As a programming library, OpenCV also offers many computer vision algorithms that will allow us to analyze either one image or a series of images obtained by the rover's camera. OpenCV recognizes patterns of pixels and colors to identify edges, objects, people, faces, etc. Our rover will use the edge/object-detection algorithms to determine if there are objects in front that the LiDAR did not detect.

OpenCV

OpenCV works using a very cool engineering concept called pipelines. To illustrate the pipeline concept, imagine a water treatment plant. Water coming into the plant may be contaminated (rainwater, sewage, etc.). It sends the water through a filter that removes large items (leaves, sticks, fish, etc.) and outputs WATER. It is still water, just a little cleaner. This cleaner water proceeds through a second filter that removes smaller objects (sand, sediment, etc.) and outputs WATER. This more sterile water is then sent through a third filter and so on until the quality of the water is at the desired level. The maintenance is easy—if one filter needs to be changed, change it. No other filters need to be changed. If we find that one filter is not working the way we want it to, we need to update it.

This process is how OpenCV works with images and "filters." Simple functions (called filters) process the images. After the filter is done, a new altered image is returned (see Figure 8-1) and can be the input to a new filter. The filters do simple image manipulation, such as sharpen, fuzz, and edge detect. Sharpen will emphasize contrast in an image, while fuzz will do the opposite. However, applying these two filters one after the other will not give us the original image, but rather a very similar image with random artifacts removed; i.e., bad pixels in a camera's image signal processor (ISP). The edge-detection filters can be general or specific; i.e., identify vertical or horizontal lines (two different filters).

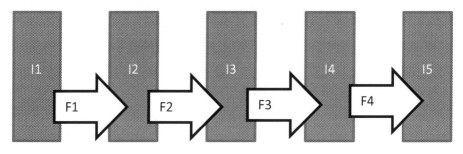

Figure 8-1. *Pipeline structure*

An image is passed into a function, which returns a modified image. The modified image can then be sent to the following process, and so on. Changing the filter/function F3 to T3' (Figure 8-2) transforms the image into the next pipeline stage. The filters/functions "downstream" do not have to change, but the next and final images (I4' & I5') will be different because of the change of the filter (F3').

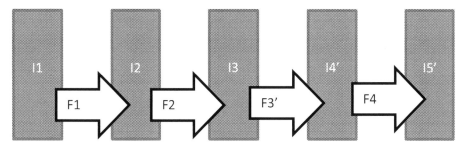

Figure 8-2. *Change filter in the pipeline (take 1)*

Furthermore, you can change the order of operations of the functions and get yet another final image. (See Figure 8-3.) In this example, we perform function F4 before F3 and get a whole new final image (I5'). This pipeline concept separates the function from the form. The form will always be an image, and the process will always be an image filter. The image filter can be designed and programmed in isolation since it does not need to know previous filters. It just "knows" an image is coming in, and it will "spit out" a "changed" image.

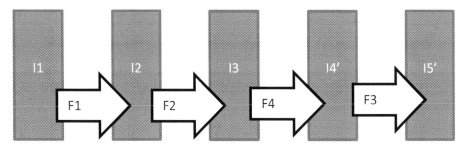

Figure 8-3. *Change filter in the pipeline (take 2)*

Images

A printed color image is a set of many characteristics, such as color and shape, interpreted by the human to be a visual representation of the environment. A digital image is a 2D array of numbers programmatically converted into a visual representation. Each array element is a 32-bit integer called a pixel, and each pixel contains four 8-bit "channels" describing the color and transparency (RGBA) at that location on the image. RGBA stands for Red, Green, Blue, and Alpha, where *Alpha* means opacity. Each 8-bit channel gives us 256 shades of color or opacity (antonym of transparency). This description involves a lot of information! We need to process this image to find edges of objects that might be in the path of the rover.

To simplify our image, the first thing to admit is that the environment has no transparent objects! After all, we are in an Egyptian pyramid. So we can ignore the Alpha channel. The second thing to observe is that the inside of an old pyramid will not have lots of colors. We can compress our image into a grayscale; i.e., only shades of gray. This first image processing leads us to our first filter.

Filters

Filters change one representation of an object into another by adding or removing information. For instance, a pair of sunglasses removes harsh reflected light from entering our eyes. What we see is still present; it just looks different. The filter removal of the harsh light gave us a simpler version of our environment.

The first filter we apply converts the RGB into gray, taking 24 bits of data and compressing it to 8 bits. We will therefore need to review color filters.

Color Filters and Grayscale

A color space is a three-dimensional computational model that mimics human visual perception known as color, where the coordinates of the model will define a perceived color. One color-space model is the RGB model, where all the colors are created by mixing red, green, and blue. Edge detection in RGB color space is difficult, because the color space is non-linear. Luckily, OpenCV has an RGB-to-Grayscale conversion function, cvtColor(image, cv2.COLOR_BGR2GRAY), that simplifies edge detection by converting the 24-bit image to an 8-bit image. (See Figure 8-4.)

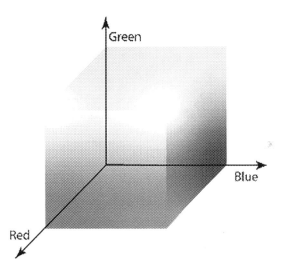

Figure 8-4. *The RGB color space conversion to grayscale space*

Image filtering will first require us to detect specific colors in images. Our image filtering will be based on a color-space model known as HSV (hue saturation value). This color-space model closely models how humans perceive colors. The HSV model is a non-linear model of RGB with cylindrical coordinates. The cylindrical coordinates allow us to graphically represent the hue, saturation, and value parameters.

Most digital color analysis programs use the HSV scale, and HSV color models are beneficial for selecting precise colors for image processing. The HSV scale provides a numerical readout of your image corresponding to the color names. Hue ranges in degrees from 0 to 360. For instance, cyan falls between 181 and 240 degrees, and magenta falls between 301 and 360 degrees—the analysis of value and saturation of color range on a 0 to 100 percent scale (Figure 8-5).

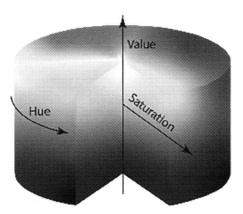

Figure 8-5. *The HSV color-space model*

We will now apply a simple color filter to our camera stream for the rover. We will identify, isolate, and separate the red, green, and blue colors in that camera stream (Figure 8-6). Testing this simple color filter will be necessary for finding the boundaries of an image's colors.

Figure 8-6. *Apply a simple color filter to our camera stream*

We use the HSV color-space models to define the required colors. The color components themselves are defined by the hue channel, which has the entire chromatic spectrum present, compared to the RGB, where we need all three channels to define a color.

For better comprehension of this part, use Figure 8-7 to approximate how the colors are defined in the hue channel.

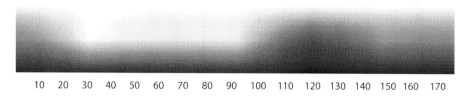

10   20   30   40   50   60   70   80   90   100   110   120   130   140   150   160   170

Figure 8-7. *Defining colors in the hue channel*

For example, if the color I'm looking for is blue, my hue range should be between 110 and 120, or 100 and 130 for a wider range. So the value of my lower limit should look something like `min_blue` = `np.array([110,Smin,Vmin])` and the higher limit like `max_blue` = `np.array([120,Smax,Vmax])`. In the case of saturation and value, we can say that the lower the saturation, the closer to white, and the lower the value, the closer to black.

337

Edge Detectors

There are two categories of edge detectors: complex and straightforward. Simple edge detectors are filters that find horizontal or vertical lines. They are fast executable code with results that may not be very helpful. Complex edge detectors may find more generic lines, but they tend to be slow. Furthermore, the generic lines may have more false edges or disjoint edges.

Edge detection looks for sharp changes in neighboring pixels in an image. We assume these changes reflect the "boundaries" between two objects in an image. Discontinuities in image brightness are likely to correspond to the following:

- Discontinuities in depth (one object is behind another, or an object is in the foreground of an image)

- Discontinuities in surface orientation (object has a "crease")

- Changes in material properties (shiny/matte or metal/cloth)

- Variations in scene illumination (shade/sunlight)

Notice that the first bullet is the only time we find a boundary between objects. (Objects are old, dusty, with mostly non-reflective surfaces. Therefore, bullets 2–4 are improbable.) Since our exploration domain is the pyramids, we can simplify our list to just the first bullet.

In the ideal case, applying an edge detector to an image will lead to a set of connected pixels that outline the boundaries of objects. Thus, applying an edge-detection algorithm to an image may significantly reduce the amount of data to be processed and filter out information that may be regarded as less relevant while preserving the important structural properties of an image. If the edge-detection step is successful, the subsequent task of interpreting the original image's contents is substantially simplified. However, it is not always possible to obtain such ideal edges from real-life images of moderate complexity.

Fragmentation or unconnected curved edges will hamper the edges of non-trivial images. These missing edge segments and false boundaries do not correspond to exciting phenomena in the image, thus complicating the subsequent task of interpreting the image data.

Edge detection is one of the fundamental steps in image processing, image analysis, image pattern recognition, and computer vision techniques.

NumPy, SciPy, OpenCV, and CV_Bridge

The OpenCV runtime libraries use the regular CPU; we need the NumPy and SciPy libraries to speed up OpenCV. NumPy uses the numeric processor on the Raspberry Pi, and SciPy includes functions not in the regular math library. Furthermore, SciPy displays graphs to the screen. If you installed the full desktop version of Noetic ROS, you have NumPy, OpenCV, and CV_Bridge installed. If not, refer to the ROS.org website to add these libraries.

Testing the OpenCV CV_Bridge

ROS has an existing program that directly accepts raw images from the camera and processes them through a callback routine called imageCallBack. This routine displays the current camera image to a window on the screen from continuously streamed camera images. Although we do not do it, OpenCV could process the images to detect objects.

To add the vision to the rover, we need to do the following steps:

- Acquire images from the RGB camera.

- Pass those images to OpenCV for further processing (smoothing, edge detecting, and segmentation).

- Filter the post-processed images to identify features, objects, walls, and lines in those images.

- Control the rover to sense and avoid those same features, obstacles, walls, and lines of objects that the LiDAR may not have detected.

We must now test our ROS and OpenCV interaction to determine if ROS can obtain images and share them with OpenCV. We can now build our bridge between ROS and OpenCV.

Okay, so now that you have a brief introduction to OpenCV and a little background on image processing, it's time to describe one of the primary tasks of computer vision: image color filtering. Color filtering is the extraction of specific color information from an image. Before that, we will see some basic operations with OpenCV so you can get acquainted with this library and understand the Python code ahead.

CV_Bridge: The Link Between OpenCV and ROS

First of all, you have to know that ROS passes images from its sensors using its own sensor_msgs/image message format, but sometimes you need to use these images with OpenCV for image processing. CV_Bridge is an ROS library that links ROS and OpenCV, converting ROS images to and from the OpenCV format. The displayed CV_Bridge link is shown in Figure 8-8.

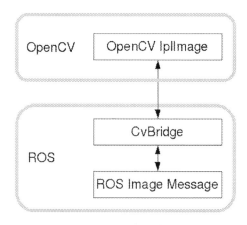

Figure 8-8. *The CV_Bridge link between ROS and OpenCV*

The following Python source-code listing allows you to call the necessary libraries to initiate the cv_bridge package and process camera messages from ROS:

```
1 from cv_bridge import CvBridge
2 oBridge = CvBridge();
3 cv_image = oBridge.imgmsg_to_cv2(image_message, desired_
encoding='passthrough');
```

On line 1, we are importing the CvBridge library. On line 2, we are instantiating the oBridge object. One line 3, we are converting the imgmsg to cv2 to send ROS sensor camera messages to OpenCV for further processing.

The following lines of code allow us to send images from ROS to be processed by OpenCV. ROS receives images from camera-sensor image message topics. That information is then sent from ROS to be further processed and analyzed by OpenCV via the CvBridge ROS package.

```
#!/usr/bin/env python

import rospy
from sensor_msgs.msg import Image
```

341

```python
from cv_bridge import CvBridge, CvBridgeError
import cv2

class ShowingImage(object):

    def __init__(self):

        self.image_sub = rospy.Subscriber("/ai_rover_
        remastered/rgb/image_raw",Image,self.camera_callback);
        self.oBridge = CvBridge();

    def camera_callback(self,data):
        try:
            # select bgr8 OpenCV encoding by default
            cv_image = self.oBridge.imgmsg_to_cv2(data,
            desired_encoding="bgr8");
        except CvBridgeError as e:
            print(e)

        cv2.imshow('image',cv_image)
        cv2.waitKey(0)

def main():
    showing_image_object = ShowingImage()
    rospy.init_node('line_following_node', anonymous=True)
    try:
        rospy.spin()
    except KeyboardInterrupt:
        print("Shutting down")
    cv2.destroyAllWindows()

if __name__ == '__main__':
    main()
```

Acquiring Test Images

Any camera images retrieved by ROS are of the sensor_msgs/image message type. To acquire those images to be further processed by OpenCV, ROS needs to subscribe to those images. The images streaming from our RGB camera located on the front of the rover (Chapter 5) need to be subscribed to by ROS nodes. Therefore, we will first develop a simple ROS node that only subscribes to the ai_rover_remastered/camera1/image_raw topic message. This topic message definition is found in Chapter 5. To build our first series of tests to determine if an ROS node can subscribe to an image and then share that image with OpenCV for further processing, we must do the following four steps:

1. To start roscore in the first terminal in Terminator:

```
$ cd ~/catkin_ws
$ source devel/setup.bash
$ catkin_make
$ roscore
```

2. We start our Gazebo simulation in a second open terminal in Terminator. The roslaunch ai_rover_remastered ai_rover_world.launch initiates our rover Gazebo simulation from Chapter 5.

```
$ cd ~/catkin_ws
$ source devel/setup.bash
$ catkin_make
$ roslaunch ai_rover_remastered ai_rover_world.launch
```

3. We start our Rviz environment in a third open terminal in Terminator. We will need to use Rviz to see our camera image. We do this to see if there are

any issues with the camera. We launch the same
Rviz launch file from Chapter 5, which is as follows:

```
$ cd ~/catkin_ws
$ source devel/setup.bash
$ catkin_make
$ roslaunch ai_rover_remastered ai_rover_remastered_rviz.launch
```

4.   We then start the fourth terminal to enter all
     required test commands.

5.   Now that we are in the fourth terminal, we want
     to run a testing command to determine if we can
     see the rover's camera. We can see it if we see ROS
     topic messages that bear the camera's topic name.
     So, if we run the following commands in the fourth
     terminal:

```
$ cd ~/catkin_ws
$ source devel/setup.bash
$ catkin_make
$ rostopic list
```

We should see the following terminal output:

```
/ai_rover_remastered/camera1/camera_info
/ai_rover_remastered/camera1/image_raw
/ai_rover_remastered/camera1/image_raw/compressed
/ai_rover_remastered/camera1/image_raw/..........
```

6.   We have confirmed that the image data is available
     on the ai_rover_remastered/camera1/image_raw
     topic. This topic message is an uncompressed image
     stream, which is more compatible with computer
     vision algorithms. Now we must write our first ROS

node that will subscribe to this image_raw topic
message. We develop this ROS node by first creating
an ROS package (cv_tests), with its associated
dependencies, for this ROS node to reside in.
We now open a new and separate terminal in
Terminator to enter the following commands:

```
$ cd ~/catkin_ws
$ source devel/setup.bash
$ cd ~/catkin_ws/src
$ catkin_create_pkg cv_tests image_transport cv_bridge sensor_
msgs rospy roscpp std_msgs
$ cd ~/catkin_ws/src/cv_tests
$ mkdir launch scripts
$ cd ~/catkin_ws/src/cv_tests/scripts
$ gedit imageSubscribeTest.py
```

7. We now have the Gedit editor open with
    imageSubscribeTest.py available, and we enter the
    following Python script listing:

```
#!/usr/bin/env python3
import rospy
from sensor_msgs.msg import Image

def imageCallBack(msg):
    pass

rospy.init_node('imageSubscribeTest');
image_sub = rospy.Subscriber('ai_rover_remastered/camera1/
image_raw', Image, imageCallBack);
rospy.spin();
```

8. We will now have to recompile our ROS package and run the imageSubscribeTest.py file with the following terminal commands:

```
$ cd ~/catkin_ws
$ source devel/setup.bash
$ catkin_make
$ cd ~/catkin_ws/src/cv_tests/scripts
$ chmod +rwx imageSubscribeTest.py
$ ./imageSubscribeTest.py
```

9. Once we run and execute the imageSubscribeTest. py node, we can then see if this node can be found in a list of active ROS nodes by entering the following terminal commands in a separate terminal in Terminator:

```
$ cd ~/catkin_ws
$ source devel/setup.bash
$ catkin_make
$ rosnode info imageSubscribeTest
```

10. We should then have the following listing output, with the publications, subscriptions, and services of the imageSubscribeTest.py that this ROS node instantiated. We can also see that /ai_rover_ remastered/camera1/image_raw connects to this ROS node.

```
Node [/imageSubscribeTest]
Publications:
 * /rosout [rosgraph_msgs/Log]

Subscriptions:
 * /ai_rover_remastered/camera1/image_raw [sensor_msgs/Image]
 * /clock [rosgraph_msgs/Clock]
```

```
Services:
 * /follower/get_loggers
 * /follower/set_logger_level

contacting node http://localhost:44103/ ...
Pid: 5791
Connections:
 * topic: /rosout
    * to: /rosout
    * direction: outbound (33717 - 127.0.0.1:41872) [10]
    * transport: TCPROS
 * topic: /clock
    * to: /gazebo (http://localhost:34451/)
    * direction: inbound
    * transport: TCPROS
 * topic: /ai_rover_remastered/camera1/image_raw
    * to: /gazebo (http://localhost:34451/)
    * direction: inbound
    * transport: TCPROS
```

11. We are confident that our ROS node connects to the camera topic message. Now we must process these images from the camera. We must pass them to the OpenCV library. This library contains real-time computer vision algorithms. To exchange images between this ROS node and OpenCV is to use the cv_bridge package that allows for this interaction. The cv_bridge package can be found in the vision_opencv stack. The following is a display of the cv_bridge acting as a link between ROS and OpenCV (Figure 8-9).

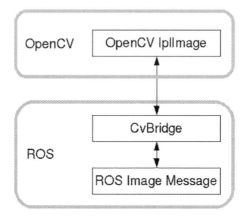

Figure 8-9. *ROS and OpenCV interaction through cv_bridge*

Please refer to the following hyperlink address for more information regarding converting ROS images and OpenCV images (Python): `http://wiki.ros.org/cv_bridge/Tutorials/ConvertingBetweenROSImagesAndOpenCVImagesPythonf`

12.   Now that we can see that a ROS node can receive an image, we need to process those images. We must test if our ROS node can send images to OpenCV through the CV_Bridge. The `cv_bridge` package contains functions that convert ROS `sensor_msgs/`Image messages to OpenCV. We must develop a Python script that converts incoming images to OpenCV messages and displays them on the OpenCV `imshow()` function. The following is a script that commands our ROS node to send images to OpenCV:

```
#!/usr/bin/env python3
import rospy
from sensor_msgs.msg import Image
import cv2, cv_bridge
```

```
class ImageSubscribeTest:

    def __init__(self):
        self.CompVisBridge = cv_bridge.CvBridge();
        cv2.namedWindow('window', 1);
        self.image_sub = rospy.Subscriber('/ai_rover_
        remastered/camera1/image_raw', Image, self.
        imageCallBack);

      def imageCallBack(self, msg):

          image = self.CompVisBridge.imgmsg_to_cv2
          (msg, desired_encoding='bgr8');
          cv2.imshow('window', image);
          cv2.waitKey(3);

rospy.init_node('ImageSubscribeTest');
imageSubscribeTest2 = ImageSubscribeTest();
rospy.spin();
```

13. Again, we wrote this testing script into the
     ImageSubscribeTest.py file in the same method
     we wrote the original file. We enter this as another
     testing script file for determining if the images
     received by ROS can be sent to the OpenCV. We
     open three terminals. We execute the roscore
     command in the first terminal, the second terminal
     runs the roslaunch ai_rover_remastered
     ai_rover_world.launch control, and the
     third terminal runs the rosrun cv_tests
     ImageSubscribeTest.py command to run the script
     that we enter in Listing 8-1. Once we run this script,
     we should see the following output (Figure 8-10).

Listing 8-1. ImageSubscribeTest.py script

```
#!/usr/bin/env python3
 import rospy from sensor_msgs.msg
 import Image import cv2, cv_bridge

class ImageSubscribeTest:
      def __init__(self):
            self.CompVisBridge = cv_bridge.CvBridge();
            cv2.namedWindow('window', 1);
            self.image_sub = rospy.Subscriber
            ('/ai_rover_ remastered/camera1/image_raw',
            Image, self. imageCallBack);
      def imageCallBack(self, msg):
            image = self.CompVisBridge.imgmsg_to_cv2
            (msg, desired_encoding='bgr8');
            cv2.imshow('window', image);
            cv2.waitKey(3);
rospy.init_node('ImageSubscribeTest');
 imageSubscribeTest2 = ImageSubscribeTest();
rospy.spin();
```

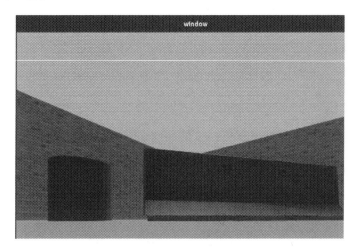

Figure 8-10. *The OpenCV displayed image*

Edge Detection and LiDAR (Why) Implementation (How)

Edge detection plays a key role during object identification and extraction, which is necessary for object avoidance. The edges detected from camera images have high vertical accuracy and represent various edge shapes of objects in the environment. But the edge detection in most camera images is affected by contrast and lighting. LiDAR data are suitable for judging building regions but miss some edge points due to the laser pulse errors. A new adaptive method of building edge detection combines LiDAR data and images to take advantage of the two data sources. First, the objects and ground are separated by a filter gradient. The non-building objects are removed by mathematical morphology and region growth. Second, the images are smoothed by Gaussian convolution, and the gradients of the image are calculated. Finally, the edge detections process the image space by the edge points of the individual roof patch. The pixels with the maximal local gradient in the buffer area are judged as the candidate edge. The ultimate edges are determined by fusing the image edges and the roof patch by morphological operation. The experimental results show that the method is adaptive for various building shapes. The ultimate edges are closed and thin with one-pixel width, which are suitable for subsequent building modeling.

Launching Python Files

This approach is a common use case in robotics: you get some data from a sensor, and you need to pass it through several parts of your applications. Each part needs the data to do its own thing, and some parts may also modify the data for other parts.

For this example, we'll create three nodes:

Node 1 (pipeline_step_1): create a random float number between 0 and 10 and publish it

Node 2 (pipeline_step_2): get this number, multiply it by 2, and publish it

Node 3 (pipeline_step_3): get this number, round it, and publish it

Figure 8-11 shows the graph we'll get at the end:

Figure 8-11. *Completed graph*

Step 1: Data Pipeline

We'll create a random float number between 0 and 10 in this node and publish it on the "data_1" topic. Here, I chose to write the node in Python:

```
#!/usr/bin/env python3
import rclpy
from rclpy.node import Node
from example_interfaces.msg import Float64
import random
class Node1(Node):
def __init__(self):
super().__init__("pipeline_step_1")
self.pub_ = self.create_publisher(Float64, "data_1", 10)
self.timer_ = self.create_timer(1.0, self.publish_data)
def publish_data(self):
msg = Float64()
msg.data = random.uniform(0.0, 10.0)
self.get_logger().info("Published: " + str(msg.data))
self.pub_.publish(msg)
def main(args=None):
rclpy.init(args=args)
```

```
node = Node1()
rclpy.spin(node)
rclpy.shutdown()
if __name__ == "__main__":Figure
main()
```

If you're not confident with writing this code, many online references (ROS.org) exist to create a minimal ROS node in Python and write a ROS Python publisher.

Here we use random.uniform() to get the required random float number.

We publish on the "data_1" topic every 1 second at 1 Hz with the timer we create.

And we also print on the terminal what we've just published, so it will be easier to debug.

Step 2: Data Pipeline

This node will subscribe to the "data_1" topic, process/transform the data, and publish the new data to the "data_2" topic. Here, I'm writing the node in C++ because ROS communications are language agnostic, so you can use any language you want for your nodes.

```cpp
#include "rclcpp/rclcpp.hpp"
#include "example_interfaces/msg/float64.hpp"
class Node2: public rclcpp::Node
{
public:
Node2() : Node("pipeline_step_2")
{
pub_ = this->create_publisher<example_interfaces::msg::Float64>
("data_2", 10);
```

```
sub_ = this->create_subscription<example_
interfaces::msg::Float64>(
"data_1", 10, std::bind(&Node2::callbackData, this,
std::placeholders::_1));
}
private:
void callbackData(const example_interfaces::msg::Float64::
SharedPtr msg)
{
auto new_msg = example_interfaces::msg::Float64();
new_msg.data = msg->data * 2.0;
RCLCPP_INFO(this->get_logger(), "Received: %lf, Published:
%lf", msg->data, new_msg.data);
pub_->publish(new_msg);
}
rclcpp::Publisher<example_interfaces::msg::Float64>::
SharedPtr pub_;
rclcpp::Subscription<example_interfaces::msg::Float64>::
SharedPtr sub_;
};
int main(int argc, char **argv)
{
rclcpp::init(argc, argv);
auto node = std::make_shared<Node2>();
rclcpp::spin(node);
rclcpp::shutdown();
return 0;
}
```

If you're not that confident with this code, check out how to write a minimal ROS node in C++.

Here, we create a publisher (to "data_2") as well as a subscriber (to "data_1").

In the "data_1" topic callback, we do the following:

- Process the data and transform it here by multiplying it by 2.

- Create a new Float64 message and fill it with this new data.

- Publish the data to the "data_2" topic.

- Print what we received and published to make debugging/monitoring easier.

Note that we didn't create any rate to publish on the "data_2" topic; we directly publish from the callback function of the "data_1" topic.

Step 3: Data Pipeline

The final node of our pipeline: this one will subscribe to the "data_2" topic, process/transform the data, and publish the new data to the "data_3" topic. And I'll write this node in Python again:

```python
#!/usr/bin/env python3
import rclpy
from rclpy.node import Node
from example_interfaces.msg import Float64
from example_interfaces.msg import Int64
class Node3(Node):
def __init__(self):
super().__init__("pipeline_step_3")
self.pub_ = self.create_publisher(Int64, "data_3", 10)
self.sub_ = self.create_subscription(
Float64, "data_2", self.callback_data, 10)
```

```
def callback_data(self, msg):
new_msg = Int64()
new_msg.data = round(msg.data)
self.get_logger().info("Received: " + str(msg.data) +
", Published: " + str(new_msg.data))
self.pub_.publish(new_msg)
def main(args=None):
rclpy.init(args=args)
node = Node3()
rclpy.spin(node)
rclpy.shutdown()
if __name__ == "__main__":
main()
```

So, we have a publisher to "data_3" and a subscriber to "data_2".
Here again, we do everything in the "data_2" callback function:

- Process the data and modify it: we round it and get an integer instead of a float number.

- Create a new message with a different type. We received Float64 data, and now we're publishing an Int64.

- Publish the new data.

- Log.

The most crucial point is that we're using a different data type to pass the message to the next pipeline step.

Note   As this is the last step of our pipeline example, we could print the result with a log and not publish it. With your nodes and data, you'll have a good idea of what you should do in your application.

Building and Running the ROS Data Pipeline Application

Add executables in `setup.py` for Python nodes and `CMakeLists. txt` for C++ nodes. Then compile from your ROS workspace with `catkin_make build`.

Running the App in Three Different Terminals

Open three terminals/sessions. If you already have three open terminals, make sure to source your ROS workspace in each one before continuing.

Let's start all three nodes and see what we get.

Terminal 1:

```
$ ROS run ROS_tutorials_py node_1
...
[INFO] [1594190744.946266831] [pipeline_step_1]: Published:
7.441288072582843
[INFO] [1594190745.950362887] [pipeline_step_1]: Published:
9.968039333074264
[INFO] [1594190746.944258673] [pipeline_step_1]: Published:
8.848880026052129
[INFO] [1594190747.945936611] [pipeline_step_1]: Published:
1.8232649263149414
...
```

Terminal 2:

```
$ ROS run ROS_tutorials_cpp node_2
...
[INFO] [1594190744.944747967] [pipeline_step_2]: Received:
7.441288, Published: 14.882576
```

```
[INFO] [1594190745.949126317] [pipeline_step_2]: Received:
9.968039, Published: 19.936079
[INFO] [1594190746.943588170] [pipeline_step_2]: Received:
8.848880, Published: 17.697760
[INFO] [1594190747.944386405] [pipeline_step_2]: Received:
1.823265, Published: 3.646530
...
```

Terminal 3:

```
$ ROS run ROS_tutorials_py node_3
...
[INFO] [1594190744.955152638] [pipeline_step_3]: Received:
14.882576145165686, Published: 15
[INFO] [1594190745.951406026] [pipeline_step_3]: Received:
19.93607866614853, Published: 20
[INFO] [1594190746.944646754] [pipeline_step_3]: Received:
17.697760052104258, Published: 18
[INFO] [1594190747.946918714] [pipeline_step_3]: Received:
3.646529852629883, Published: 4
...
```

Thanks to the logs, you can see where the data goes, when it is received and sent, and how it is processed.

Now, if you get the list of all topics running in your graph with ROS topic list:

```
$ ROS topic list
/data_1
/data_2
/data_3
/parameter_events
/rosout
```

We find the topics "data_1", "data_2," and "data_3".

So, this is great. From there, you can do the following:

- Listen to any topic with ROS topic echo from the terminal and see what's going on.

- Plug any new node into any of those topics. For example, you want to make a more complex data pipeline: another node can subscribe to "data_2" and then process it independently.

Starting Your Data Pipeline with a ROS Launch File

For developing and debugging, starting your nodes from the terminal is fantastic. However, if you want to create an actual ROS application, you'll have to use launch files. On top of the advantages of using an ROS launch file, you'll also be able to start all your nodes at the same time, so all the pipeline steps will start working simultaneously.

So, let's write a simple launch file to start all three nodes:

```
from launch import LaunchDescription
from launch_ros.actions import Node
def generate_launch_description():
ld = LaunchDescription()
node_1 = Node(
package="ROS_tutorials_py",
executable="node_1"
)
node_2 = Node(
package="ROS_tutorials_cpp",
executable="node_2"
)
```

```
node_3 = Node(
package="ROS_tutorials_py",
executable="node_3"
)
ld.add_action(node_1)
ld.add_action(node_2)
ld.add_action(node_3)
return ld
```

Now compile and run this launch file. Here's an ROS launch file tutorial if you don't know how.

```
$ ROS launch my_robot_bringup data_pipeline.launch.py
[INFO] [launch]: All log files can be found below /home/ed/.
ros/log/2020-07-08-09-55-37-919871-ed-vm-11593
[INFO] [launch]: Default logging verbosity is set to INFO
[INFO] [node_1-1]: process started with pid [11595]
[INFO] [node_2-2]: process started with pid [11597]
[INFO] [node_3-3]: process started with pid [11599]
[node_1-1] [INFO] [1594194939.304693502] [pipeline_step_1]:
Published: 0.9131821923863725
[node_2-2] [INFO] [1594194939.305641498] [pipeline_step_2]:
Received: 0.913182, Published: 1.826364
[node_3-3] [INFO] [1594194939.314622095] [pipeline_step_3]:
Received: 1.826364384772745, Published: 2
[node_1-1] [INFO] [1594194940.292584583] [pipeline_step_1]:
Published: 6.724756051691409
[node_2-2] [INFO] [1594194940.293238316] [pipeline_step_2]:
Received: 6.724756, Published: 13.449512
[node_3-3] [INFO] [1594194940.294448648] [pipeline_step_3]:
Received: 13.449512103382817, Published: 13
```

And your data pipeline is now fully ready! For production you may remove the info logs and only print the warn logs.

Summary

You can now build a complete data pipeline using ROS nodes and topics. The ROS architecture and tools bring you many advantages. You can do the following:

- Write any step or pipeline (node) in any language you want

- Debug each step from the terminal with the command-line tool

- Start only a few steps of your pipeline

- Add new steps (nodes) at the beginning, end, or anywhere else in the pipeline

CHAPTER 9

Reinforced Learning

In the previous chapter, our rover needed the ability to extract information from images and recognize objects by their patterns and features. This chapter will outline a novel framework of autonomous aviation with the application of artificial intelligence in the form of a reinforcement learning agent. This agent learns flying skills by observing a pilot's psychological reaction and flight path in a flight simulator. The framework consists of a gaming module that works as a flight simulator, a computer vision system that detects the pilot's gestures, a flight dynamics analyzer for verifying the safety limits of the state space variables (the performance of the aircraft) during a simulated flight, and a module to calculate the Q-function and the learned policy.

REL Primer

This chapter is diverse in that reinforcement learning (REL) is not supervised or unsupervised learning. There exists an agent that can observe the environment, select and perform an action/s, and get rewards in return (or penalties in the form of negative rewards). Then it must learn by itself based on what may or may not be initially given to it. This is referred to as the best strategy, called a policy, to get the most reward over time (if not given to it, it makes a strategy, or it makes the one it is given better). A policy defines what action the agent should choose when it is in

D. A. Blubaugh et al., *Intelligent Autonomous Drones with Cognitive Deep Learning*,
https://doi.org/10.1007/978-1-4842-6803-2_9

a given situation. For example, many robots implement REL algorithms to learn how to walk. DeepMind's AlphaGo program is a good example of reinforcement learning: in May 2017 it beat the world champion Ke Jie at the game of Go. It learned its winning policy by analyzing millions of games, and then playing many games against itself. The vast majority of available data is actually unlabeled: we have the input features X, but we do not have the labels Y. There is huge potential in reinforcement learning, as we have only barely scratched the service.

The central concept of REL is the notion of an intangible action based on a policy. Next is the agent's directly affecting the environment, the basic actions of RL; as the agent learns, the internal state of the agent is modified in order to build upon policy.

Semi-Markov decision processes (SMDPs) have potential benefits in REL. When a complex task is created, the Q value for the overall task can be formed by combining the Q values of the simpler tasks. This is a way of solving the problem of finding the optimal decision thresholds by casting it as a stochastic optimal control problem by choosing actions in each state in the corresponding SMDP such that the average reward rate is maximized.

A basic observation is that actions take time to complete. When an agent selects a basic action, the result of that action can be observed in the next time step (as defined by the MDPs). But actions are not completed in a single time step, as there is some time interval that elapses while the policy is executing the underlying basic actions, and only at the end of that delay period can the results of that action be observed.

SMDPs extend the basic MDP framework by adding time into the various components, thus allowing them to capture this style of decision problem. SMDPs extend the MDP framework in three ways, as follows:

- First, the transition function, P, must incorporate time. This is expressed as $P : S \times A \times /\!/x\, S \rightarrow \mathbb{R}$, where $P(s; a; t; s')$ indicates the probability of arriving in state s', t time steps after choosing action a in state s. This is a discrete time SMDP, as time is expressed as a number of time steps.

Continuous time SMDPs also exist, where $P : S \times A \times /\!/x$ $S \to \mathbb{R}$, but in general we will describe things here in the simpler discrete time case.

- Next, the reward for an action a is no longer a single value given when the action is chosen, but instead represents the total reward accumulated from when the action is chosen to when the next state is reached. In terms of sub-tasks, that means that a chosen sub-policy could execute several actions, some or all of which receive some reward, and the total reward for choosing that sub-policy is the sum of those rewards.

- The final change is in the discounting factor, γ. This is now applied across the time delay, so that states that are not reached for a long time are discounted more than states that are reached quickly.

To illustrate all this, we'll present an example case study using an REL.

REL in Aviation Case Study: unmanned or manned, with an injection of AI. In this example (Krishnamurthy, Harbour, & Clark 2019), the pilot and the REL agent learn flying skills simultaneously, forming a symbiotic relationship. The episodes for training the reinforcement learning agent can be simulated by a pilot flying in a simulator, or unmanned using a game on a computer. In a typical episode, the reinforcement learning agent provides a sequence of actions for the pilot to follow.

These instructions produce one of the two types of state, either success or failure. The agent observes the psychological reaction of the pilot as well as the flight environment and receives a positive or a negative reward. The trained REL agent represents a novel form of AI that guides the pilot for various phases of flight.

Human error is causal to most aircraft accidents; consequently, technologies have emerged to issue warnings when the aircraft's travel trajectory is irregular. For example, detecting the aircraft's behavior is one way to measure the safety of the aircraft.

Continuous monitoring and analysis of flight operations is another approach to detect hazardous behavior from a pre-defined list. Li et al. (2016) have reported data-mining methods such as cluster analysis of digital flight data using Gaussian mixture models (GMMs) that are employed by safety analysts to identify unusual data patterns or anomalies and latent risks in daily operations. With the advent of artificial intelligence (AI), human-autonomy teaming can be an efficient way to minimize human error and further increase aviation safety records.

Zhao et al. (2018) have used reinforcement learning (REL) as an adaptive online learning model to identify common patterns in flight data and to update the clusters for GMM using a recursive expectation-maximization algorithm. The resurgence of interest in artificial intelligence has attracted applications in aviation systems, in particular, air-traffic management (ATM), air traffic flow management (ATFM), and unmanned aerial systems traffic management (UTM).

Kistan et al. (2018) have explored a cognitive human-machine interface (HMI), configured via machine learning, and examined the requirements. They postulated that increased automation and autonomy through AI will lead to certification requirements and discussed how ground-based ATM systems can be accommodated by the existing certification framework for aviation systems. The recent developments in AI open up possibilities in autonomous aviation for introducing a high level of safety by replacing a pilot's actions with robotic functions, and further research on how AI can be incorporated into autonomous aviation is highly desirable.

Pilot-modeling technologies have played a crucial role in manned aviation, and control models of human pilot behavior have been developed. Control models are used to analyze the characteristics of the pilot–aircraft system for guidance in the flight-control system. Anthropomorphic models of human operators, which cover the central nervous system, neuromuscular system, visual system, and the vestibular system, can represent a pilot's behavior.

Recently, Xu et al. (2017) have reviewed control models of human pilot behavior. These models reflect the dynamics of a human sensory and control effector that acts in response to a stimulus. AI in the form of computer vision can be coupled with these models to detect non-linear characteristics of human pilot behavior for training the REL agent.

Simulators for Emotion Recognition

Flight simulators are routinely used both for training pilots and for evaluating war fighters' preparedness for missions. Flight simulators have also been used in research studies on the relationship between emotional intelligence and simulated flight performance to understand how emotional factors affect flight-training performance. Pour et al. (2018) have used a human-robot facial expression reciprocal interaction platform to study the social interaction abilities of children with autism.

Reinforcement Deep Learning

Reinforcement learning is a type of semi-supervised learning inspired by the way animals learn, by way of being punished when things or the process or the choice made is incorrect or ends badly or being rewarded when things or the process or the choice made is correct or ends well. It relies on the definition of state space, actions for transitions between states, and an associated reward structure in a Markov decision process. In some of the simple forms of REL, one learns an optimal policy by evaluating the value functions $V(s)$ or $Q(s,a)$-learning, where s is the current state and a is the action taken in state (s), or temporal differences from episodes, which are example attempts by agents to reach the goal. In a game setting, the episodes can be either successful or unsuccessful attempts. The game-like situations are realized in many daily-life examples, including pilot attempts to fly an aircraft. The optimal policy is used to select action sequences to transition between states and to achieve the final goal while maximizing the long-term rewards.

Computer Vision System

It is also possible to design a computer vision system to capture the psychological reaction of a pilot undergoing training in a simulator. This helps ascertain the pilot's reaction on a flight path following the pilot's operational action. To train the REL agent, we design a flight simulator framework, which is like a game for the pilot to play using his or her actions, *a*, and expressing his or her gestures, which are representative of his or her actions in the simulator. We represent the gesture, *g*, as a two-state variable, with values *happy* (☺) or *unhappy* (☹). The state space of the flight simulator, *s*, consists of five variables: altitude, *A*, speed, *S*, heading, *H*, turn, *U*, and roll, *R*. Table 9-1 lists the range of these five action variables.

Table 9-1. *The Five Variables That Define the State s and Their Ranges*

State Variable	Minimum	Maximum
Altitude, A	0 ft	35,0000 ft
Speed, S	0 mph	550 mph
Heading, H	0°	360°
Turn, U	0°	360°
Roll, R	0°	360°

Flight-Path Analysis

A reliable flight-path analysis can be obtained by real-time computation of the gradients of state space variables, namely altitude gradient dA/dt, speed gradient dS/dt, heading gradient dH/dt, turn gradient dU/dt, and roll gradient dR/dt. A rule-based model compares the gradients with a predefined range to determine if the maneuver is safe or risky and calculate a dynamic reward. Table 9-2 shows the gradients and the initial guess values of their ranges. The minimum and maximum of the range can be set as tunable parameters so that a more practical set of values is achieved iteratively. These ranges will be used in a rule model to dynamically determine the reward for the flight maneuvers.

Table 9-2. *Ranges of Gradients of the State Variables That Define the Safe Operational Zone*

Gradient of State Variable	Minimum	Maximum
Altitude gradient, dA/dt	0 ft	1,000 ft
Speed gradient, dS/dt	0 mph	20 mph
Heading gradient, dH/dt	0°	3°
Turn gradient, dU/dt	0°	3°
Roll gradient, dR/dt	0°	2°

Pilot's Gesture Assignment

A computer vision system can consist of a digital video camera, a neural processing unit such as Myriad 2, and a single board computer to read the pilot's gestures. The computer vision system can be trained using a face-detection machine learning algorithm for real-time monitoring of the "happy" or "unhappy" facial expression of the pilot. The agents who play the game of flying the plane in the simulator are particularly advised to show a happy gesture (☺) when their actions result in a safe operation and an unhappy gesture (☹) when their actions result in a risky, unsafe, or catastrophic operation. The computer vision system can be as simple as a Google AIY kit, which operates with a TensorFlow machine learning model to detect a smile.

Reinforcement Learning Agent: Learning from Pilot's Actions

We first consider a human-in-the-loop approach to develop an REL agent that can use artificial intelligence to determine a human pilot's gesture and calculate rewards. This type of reinforcement learning agent is trained with episodes that are generated when a pilot is flying an aircraft in a simulator; i.e., on a computer. In a typical episode, the REL agent provides a sequence of actions for the pilot to follow. These instructions produce one of two results, either success or failure. The agent observes the psychological reaction of the pilot as well the flight environment and receives a positive or negative reward. The agent receives a +1 reward when the pilot's gesture is "happy" and a -1 reward when the pilot's gesture is "unhappy." For example, the same set of episodes can be trained with different reward structures to find out an optimal set for safe operations. The training process is repeated until convergence of the learning process is achieved. After training the agent using a sufficiently large number of episodes, the knowledge acquired by the REL agent is expected to represent a novel form of AI that directs the pilot with accurate instructions for various phases of flight. Figure 9-1 shows a learning framework of an REL agent and its interactions with the flight simulator and the computer vision system that detects the pilot's gesture for receiving rewards to update the $Q(s,a)$ function, where s is the current state and a is the action, and the policy is $\pi(s,a)$.

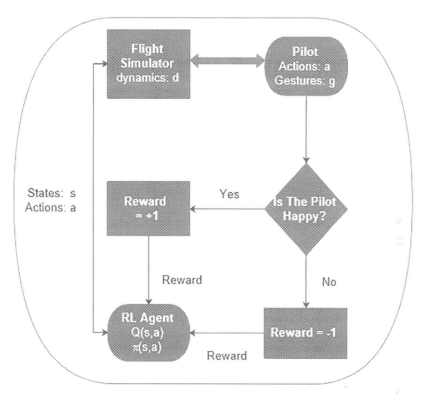

Figure 9-1. *A framework for the reinforcement learning (REL) agent and its interactions with its environment, consisting of the flight simulator and the pilot's gesture-recognition system*

Flight Simulator Game Framework

Figure 9-2 shows a game framework in a flight simulator for generating the states, actions, rewards, Q function, and policy. The flight simulator game framework has an additional local reward and a long-term reward compared to the reward structure of the REL agent. The game REL agent in the flight simulator game framework receives as additional reward of -1 for each instance of a state variable's gradient falling outside the safe range. An optional long-term reward of +2 is also awarded to the game REL agent when the total time taken to reach the destination is below a preset value. The REL agent will receive a reward of +1 when all of the state variable's derivatives are within the safe range. The choice of rewards is arbitrary and can evolve to a more realistic one based on various episodes. A game simulator module initiates the game by extracting actions, using the current policy to simulate the flight dynamics. Then, two other modules evaluate the flight dynamics and the gesture of the pilot to identify the rewards. Then, the $Q(s,a)$ function is calculated and updated for each state–action pair and the associated reward. The policy $\pi(s,a)$ is then recalculated from the $Q(s,a)$ values and updated.

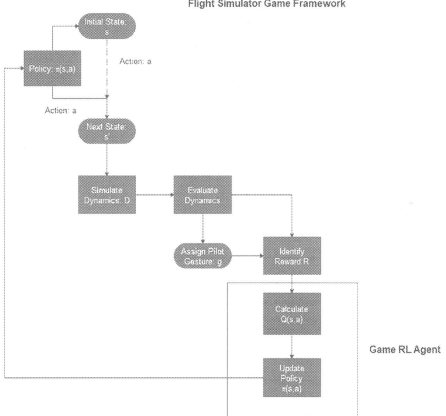

Figure 9-2. *The game framework for flight simulator as a game for obtaining the states, actions, rewards, Q function, and policy*

Summary

This chapter presented a novel theme of aviation with the injection of AI in the form of an REL agent that learns flying skills by observing the pilot's psychological reaction and flight path in a simulator. A unique feature of this AI framework is that the pilot and the REL agent learn flying skills simultaneously, forming a symbiotic relationship.

In this context, identifying suitable methods for detecting pilot behavior is key to developing an AI based on reinforcement learning. With sufficient training within the proposed framework, the REL agent is expected to learn to fly the aircraft as well as to guide the pilot for safe aviation.

Policies and Value Functions

- Policy map states to actions

  - Probability of taking - action a while in state s

    - $\pi(a|s) = \mathbb{P}[A_t = a| S_t = s]$

    - Value function

  - Provides reward for action based on agent's state while following the policy

    - $V_\pi(s) = \mathbb{E}_\pi[G_t|S_t = s]$ (measures how good state is)

    - $Q_\pi(s,a) = \mathbb{E}_\pi[G_t|S_t = s|, A_t = a]$ (measures how good it is to take action)

    - Estimate new Q-value if using tabular Q-learning:

$$New\, Q(S, A) \leftarrow Q(S, A) + \alpha\left[R' + \gamma \max_a Q'(S', a) - Q(S, A)\right]$$

  - Return function

    - $G_t = \sum_{k=0}^{\infty} \gamma^k r_{t+k+1} | S_t = s$

References

Chang, T.H., Hsu, C.S., Wang, C., & Yang, L.-K. (2008). On board measurement and warning module for measurement and irregular behavior. *IEEE Transactions on Intelligent Transportation Systems*, 9(3), 501–513.

Clark, J.D., & Harbour, S.D. (2019). Unpublished.

Clark, J.D., Mitchell, W.D., Vemuru, K.V., & Harbour, S.D. (2019). Unpublished.

Dayan, P., & Abbott, L. F. (2001). *Theoretical neuroscience: computational and mathematical modeling of neural systems*. MIT Press

Gerstner, W., & Kistler, W. (2002). *Spiking Neuron Models: Single Neurons, Populations, Plasticity*. Cambridge University Press.

Friston, K., & Buzsáki, G. (2016). The functional anatomy of time: what and when in the brain. *Trends in cognitive sciences*, 20(7), 500–511.

Friston K. (2018). Am I Self-Conscious? (Or Does Self-Organization Entail Self-Consciousness?). *Frontiers in psychology*, 9, 579. doi:10.3389/fpsyg.2018.00579

Harbour, S.D., & Christensen, J.C. (2015, May). A neuroergonomic quasi-experiment: Predictors of situation awareness. In: *Display Technologies and Applications for Defense, Security, and Avionics IX; and Head-and Helmet-Mounted Displays XX* (Vol. 9470, p. 94700G). SPIE.

Harbour, S.D., Clark, J.D., Mitchell, W.D., & Vemuru, K.V. (2019). Machine Awareness. *20th International Symposium on Aviation Psychology*, 480–485. https://corescholar.libraries.wright.edu/isap_2019/81

Harbour, S.D., Rogers, S.K., Christensen, J.C., & Szathmary, K.J. (2015, 2019). Theory: Solutions toward autonomy and the connection to situation awareness. Presentation at the 4th Annual Ohio UAS Conference. Convention Center, Dayton, Ohio. USAF.

Kidd, C., & Hayden, B.Y. (2015). The Psychology and Neuroscience of Curiosity. *Neuron*, 88(3), 449–460.

Kistan, T., Gardi, A., & Sabatini, R., (2018). Machine learning and cognitive ergonomics in air traffic management: Recent developments and considerations for certification. *Aerospace, 5*(4), Article Number 103.

Li, L.S., Hansman, R.J., Palacios, R., & Welsch, R. (2016). Anomaly detection via Gaussian mixture model for flight operation and safety monitoring. *Transportation Technologies, Part C: Emerging Technologies, 64,* 45–57.

Loewenstein, G. (1994). The Psychology of Curiosity: A Review and Reinterpretation. *Psychological Bulletin. 116*(1), 75–98.

Mitchell, W.D. (February, 2019). Private communication.

Murphy, R.R. (2019). *Introduction to AI robotics.* MIT press.

Pour, A.G., Taheri, A., Alemi, M., & Meghdari, A. (2018). Human-Robot facial expression reciprocal interaction platform: Case studies on children with autism. *International Journal of Social Robotics, 10*(2), 179–198.

Rogers, S. (2019). Unpublished.

Sharpee, T.O., Calhoun, A.J., & Chalasani, S.H. (2014). Information theory of adaptation in neurons, behavior, and mood. *Current opinion in neurobiology, 25,* 47–53.

Vemuru, K.V., Harbour, S.D., & Clark, J.D. (2019). Reinforcement Learning in Aviation, Either Unmanned or Manned, with an Injection of AI. *20th International Symposium on Aviation Psychology,* 492–497. https://corescholar.libraries.wright.edu/isap_2019/83

Xu, S.T., Tan, W.Q., Efremov, A.V., Sun, L.G., & Qu, X. (2017). Review of control models for human pilot behavior. *Annual Review in Control, 44,* 274–291.

Zhao, W.Z., He, F., Li, L.S., and Xiao, G. (2018). An adaptive online learning model for flight data cluster analysis, In: *Proc. of 2018 IEEE/AIAA 37th Digital Avionics Systems Conference, IEEE-AIAA Avionics Systems Conference* (pp.1–7). London, UK.

CHAPTER 10

Subsumption Cognitive Architecture

This chapter discusses subsumption cognitive architecture ideas, including approaches, potential methods, and examples, and then we finish with where the field is headed, plus other techniques. By the end of the chapter, you will have an understanding of the following:

- Subsumption cognitive architecture (SCA)

- Subsumption structure's four key areas: drone situated awareness (SA), embodiment, intelligence, and emergence

- How to program your drone to possess a subsumption cognitive architecture by utilizing the provided Python code examples

- How SCA compares to other cognitive architectures

Cognitive Architectures for Autonomy

Subsumption cognitive architecture (SCA) involves a construct that is a management structure proposed to guide conduct via symbolic mental representations of the world; subsumption architecture couples sensory

© David Allen Blubaugh, Steven D. Harbour, Benjamin Sears, Michael J. Findler 2022
D. A. Blubaugh et al., *Intelligent Autonomous Drones with Cognitive Deep Learning*,
https://doi.org/10.1007/978-1-4842-6803-2_10

data to motion determination in an intimate and bottom-up fashion. It does this by way of decomposing the whole conduct into sub-behaviors. These sub-behaviors are broken down into a hierarchy of layers. Each layer implements a precise degree of behavioral competence, and higher stages can subsume lower degrees (a.k.a., integrate/combine lower levels into a more comprehensive whole) to create possible behavior. For example, a drone's lowest layer should be "avoid an object." The second layer would be "wander around," which runs underneath the third layer, "explore the world." Because a robot needs to have the capacity to "avoid objects" to "wander around" effectively, the subsumption architecture creates a device in which the higher layers use the lower-level competencies. The layers, which all acquire sensor information, work in parallel and generate outputs. These outputs can be instructions to actuators or signals that suppress or inhibit different layers.

Subsumption Structure

Subsumption structure is a reactive robotic structure closely related to behavior-based robotics. Subsumption has been extensively influential in independent robotics and real-time AI. It utilizes intelligence from a considerably exclusive perspective. Drones are based on the whole notion of intelligence, corresponding to unconscious thinking processes. Instead of modeling factors of human genius via image manipulation, this approach is aimed at real-time interplay and possible responses to a dynamic lab or workplace environment.

The intention is to be knowledgeable using the following four key ideas:

1. Drone Situated Awareness (SA) – An essential concept of SA AI is that a robot should be in a position to react to its surroundings within a human-like time frame. The situated mobile robot is not to symbolize

the world by utilizing an inside set of symbols and then act on this model. Instead, "The actual real world is its model." a skill that a suitable perception-to-action setup can use to directly interact with the world as opposed to modeling it. Yet, every module/behavior nevertheless models the world, but close to the sensorimotor signals on an efficient level. These accessible models necessarily use assumptions about the world that are hardcoded in the algorithms themselves; however, they do avoid using memory to predict the world's behavior, instead relying on direct sensory feedback as much as possible.

2. Embodiment – Constructing a consolidated agent completes two artifacts. First, it forces the designer to check and to create an integrated bodily management system, no longer theoretical fashions or simulated robots that might no longer work in the physical world. The second is that it can clear up the image grounding problem, a philosophical difficulty many common AIs encounter, without delay, coupling sense data with significant actions. The world grounds regress, and the inner relation of the behavioral layers is directly grounded in the world the robot perceives.

3. Intelligence – Looking at evolutionary progress, developing perceptual and mobility skills is a critical basis for human-like intelligence. Also, by using top-down representations as a workable beginning point for AI, it seems that "intelligence is decided by way of the dynamics of interaction with the world."

4. Emergence – Conventionally, individual modules are not regarded as shrewd in themselves. It is the interplay of such modules, evaluated by gazing at the agent and its environment, that is generally deemed competent (or not). "Intelligence," therefore, "is in the eye of the observer."

The ideas just outlined are still a part of an ongoing debate involving the nature of the brain and how the development of robotics and AI ought to be fostered.

One idea could be to add intelligence in layers, like downloading "apps" as needed or upgrading to a newer or better version. It all comes down to how much intelligence the designer believes the robot needs (Harbour et al. 2019; Murphy 2019).

Layers and Augmented Finite-State Machines

Each layer comprises a deposit of processors that are augmented finite-state machines (AFSM), the augmentation being delivered with "instantiation" variables to keep programmable data structures. A layer is a module and is accountable for a single behavioral goal, such as "wander around." There is no central manipulator inside or between these behavioral modules. All AFSMs continually and asynchronously receive input from the applicable sensors and send output to actuators (or different AFSMs). Input alerts that are not examined before a new one is delivered wind up getting discarded. These discarded indicators are common and are beneficial for overall performance because discarding them lets the robot work in real time by dealing only with the most immediate information.

Because there is no central control, AFSMs talk with each other through inhibition and suppression signals. Inhibition indicators block indicators from accomplishing actuators or AFSMs, and suppression alerts block or replace the inputs to layers or their AFSMs. This system of AFSM verbal exchange is how better optimization layers subsume lower ones (see Figure 10-1), as properly as how the structure offers priority and motion determination arbitration in general to make the best choice. An example could be 1.) Start with Explore World. 2.) Clearly it will subsume Wander Around. 3.) Next Avoid Objects while Exploring World.

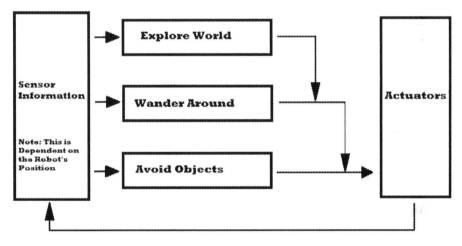

Figure 10-1. *Abstract representation of subsumption architecture, with the higher-level layers subsuming the roles of lower-level layers when the sensory information determines it.*
(Brooks, Rodney. [1999]. Cambrian Intelligence: The Early History of the New AI. Cambridge, MA: MIT Press.)

The improvement of layers follows an instinctive progression. The lowest layer is created, tested, and fixed. Once that lowest stage runs, one attaches the second layer with the proper suppression and inhibition connections to the first layer. After testing and fixing the blended behavior, this technique can be repeated for (theoretically) any variety of behavioral modules.

Examples Using a Cognitive Assumption Architecture

Note   Check out Carnegie Mellon University's open source Robot Navigation Toolkit (CARMEN) at `http://www.cs.cmu. edu/~carmen/links.html`.

Rython robotics (Pyro) is composed of a set of Python trainings that encapsulate the lower-stage details (`https://works.swarthmore.edu/ cgi/viewcontent.cgi?article=1009&context=fac-comp-sci`).

Figure 10-2 presents a schematic of the Pyro architecture. Users write robotic control packages via a single application programming interface (API). The API is carried out as an object-oriented hierarchy that provides an abstraction layer atop all the supplier-supplied precise robotic APIs. For example, in Figure 10-2, all the unique APIs have been abstracted into the classification `pyro.robot`. In addition, different abstractions and services are handy in the Pyro library. The libraries assist to simplify the robot's unique facets and to supply insulation from the important lower-stage points of the hardware or simulation environments (Blank, Kumar, Meeden, & Yanco, 2003).

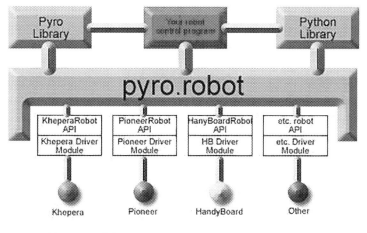

Figure 10-2. *Pyro architecture*

It is no longer necessary to recognize all the details of the Pyro implementation, but the reader should observe that the whole control program is independent of the type of robotics and the type of sensor being used. Listing 10-1 will keep the robot away from barriers when they are within the safe distance of 1 robot unit (discussed below) of the robot's front left or front center various sensors (lines 6 and 9, respectively), agnostic of the kind of robot. Lines 14 and 15 exhibit the details of Pyro's automated initialization mechanism. Such traces will be left out in subsequent examples (Blank et al. 2003).

Listing 10-1. An Obstacle Avoidance Program, in Pseudocode and in Pyro

```
# if approaching an obstacle on the left side #
 turn right
# else if approaching an obstacle on the right side #
 turn left
# else go forward
1 from pyro.brain import Brain
```

```
2 class Avoid(Brain):
3 def step(self):
4 safeDistance = 1 # in Robot Units
5 #if approaching an obstacle on the left side, turn right
6 if min(self.get('robot/range/front-left/value')) <
safeDistance:
7 self.robot.move(0,-0.3)
8 #else if approaching an obstacle on the right side, turn left
9 elif min(self.get('robot/range/front-right/value')) <
safeDistance:
10 self.robot.move(0,0.3)
11 #else go forward
12 else:
13 robot.move(0.5, 0)
14 def INIT(engine):
15 return Avoid('Avoid', engine)
```

(Pyro: A Python-based Versatile Programming Environment
For .... https://works.swarthmore.edu/cgi/viewcontent.
cgi?article=1009&context=fac-comp-sci)

Most of the Pyro framework is written in Python, an easy-to-read
scripting language that appears very similar to pseudocode. It additionally
integrates effortlessly with C and C++ code, which makes it feasible to
rapidly incorporate existing code. The C/C++ interface also enables very
costly routines (like imaginative and prescient programs) at lower tiers for
quicker runtime efficiency. Also, we are able to "wrap" applications written
in C and C++ (such as Player/Stage) so that they are instantly, and natively,
handy in Python. Refer to Listings 10-2 through 10-4, which are discussed
in the sections that follow.

Listing 10-2. A Wander Program (Blank et al. 2003)

```
from pyro.brain import Brain
from random import random class Wander(Brain):
def step(self):
safeDistance = 0.85 # in Robot Units
l = min(self.get('robot/range/front-left/value')) r = min(self.
get('robot/range/front-right/value')) f = min(self.get('robot/
range/front/value'))
if (f < safeDistance): if (random() < 0.5):
self.robot.move(0, - random()) else:
self.robot.move(0, random()) elif (l < safeDistance):
self.robot.move(0,-random()) elif (r < safeDistance):
self.robot.move(0, random()) else: # nothing blocked, go
straight
self.robot.move(0.2, 0)
(Pyro: A Python-based Versatile Programming Environment
For .... https://works.swarthmore.edu/cgi/viewcontent.
cgi?article=1009&context=fac-comp-sci)
```

Listing 10-3. A Neural Network Controller (Blank et al. 2003)

```
from pyro.brain import Brain
from pyro.brain.conx import Network class NNBrain(Brain):
def setup(self): self.net = Network()
self.net.addThreeLayers(self.get('robot/range/count'), 2, 2)
self.maxvalue = self.get('robot/range/maxvalue')
def scale(self, val):
return (val / self.maxvalue) def teacher(self):
safeDistance = 1.0
if min(self.get('robot/range/front/value')) < safeDistance:
trans = 0.0
```

```
elif min(self.get('robot/range/back/value')) < safeDistance:
trans = 1.0
else:
trans = 1.0
if min(self.get('robot/range/left/value')) < safeDistance:
rotate = 0.0
elif min(self.get('robot/range/right/value')) < safeDistance:
rotate = 1.0
else:
rotate = 0.5 return trans, rotate
def step(self):
ins = map(self.scale, self.get('robot/range/all/value'))
targets = self.teacher()
self.net.step(input = ins, output = targets)
trans = (self.net['output'].activation[0] - .5) * 2.0 rotate
= (self.net['output'].activation[1] - .5) * 2.0 robot.
move(trans, rotate)
(Pyro: A Python-Based Versatile Programming Environment
For .... https://works.swarthmore.edu/cgi/viewcontent.
cgi?article=1009&context=fac-comp-sci)
```

Listing 10-4. A Finite-State Machine Controller (Blank et al. 2003)

```
from pyro.geometry import distance
from pyro.brain.behaviors.fsm import State, FSMBrain
class edge(State):
def onActivate(self):
self.startX = self.get('robot/x')
self.startY = self.get('robot/y')
def update(self):
x = self.get('robot/x')
y = self.get('robot/y')
```

```python
dist = distance( self.startX, self.startY, x, y)
if dist > 1.0:
self.goto('turn')
else:
self.robot.move(.3, 0)
class turn(State):
def onActivate(self):
self.th = self.get('robot/th')
def update(self):
th = self.get('robot/th')
if angleAdd(th, - self.th) > 90:
self.goto('edge')
else:
self.robot.move(0, .2)
def INIT(engine):
brain = FSMBrain(engine)
brain.add(edge(1)) # 1 means initially active
brain.add(turn())
return brain
```

(Pyro: A Python-based Versatile Programming Environment
For .... https://works.swarthmore.edu/cgi/viewcontent.
cgi?article=1009&context=fac-comp-sci)

Controlling the Robotic Car

Now we'll describe how to program a Raspberry Pi that directly controls a robotic car. The robot vehicle is the same just used; however, it now uses subsumption architecture to manage the behaviors. Python is the language used for the subsumption instructions and scripts. Creating subsumptive Java classes is done with leJOS. You can study more about these Java lessons at www.lejos.org. There are two major classes required: one abstract category named `Behavior` and the other named `Controller`. The `Behavior` class encapsulates the car's conduct via the use of the following methods:

- `takeControl`: Returns a Boolean price indicating if the `Behavior` class has to take control or not

- `action`: Implements the particular conduct completed by way of the car

- `suppress`: Causes the motion conduct to immediately stop, and then returns the automobile state to one in which the subsequent conduct can occur (Norris 2019).

```
control.import RPi.GPIO as GPIO
import time
class Behavior(self):
global pwmL, pwmR
# use the BCM pin numbers
GPIO.setmode(GPIO.BCM)
# set up the motor control pins
GPIO.setup(18, GPIO.OUT)
GPIO.setup(19, GPIO.OUT)
pwmL = GPIO.PWM(18,20) # pin 18 is left wheel pwm
pwmR = GPIO.PWM(19,20) # pin 19 is right wheel pwm
# must 'start' the motors with 0 rotation speeds
pwmL.start(2.8)
pwmR.start(2.8)
(Beginning Artificial Intelligence With The Raspberry Pi ....
https://idoc.pub/documents/beginning-artificial-intelligence-
with-the-raspberry-pi-1430zjgkdo4j)
```

Controller Class and Object

The `Controller` class contains the main subsumption logic that determines which behaviors are active based on priority and the need for activation. The following are some of the methods in this class:

- `__init__()`: Initializes the `Controller` object

- `add()`: Adds a behavior to the list of available behaviors. The order in which they are added determines the behavior's priority.

- `remove()`: Removes a behavior from the list of available behaviors. Stops any running behavior if the next highest behavior overrides it.

- `update()`: Stops an old behavior and runs the new behavior

- `step()`: Finds the next active behavior and runs it

- `find_next_active_behavior()`: Finds the next behavior wishing to be active

- `find_and_set_new_active_behavior()`: Finds the next behavior wishing to be active and makes it active

- `start()`: Runs the selected action method

- `stop()`: Stops the current action

- `continously_find_new_active_behavior()`: Monitors in real-time for new behaviors desiring to be active

  (Beginning Artificial Intelligence with the Raspberry Pi .... `https://idoc.pub/documents/beginning-artificial-intelligence-with-the-raspberry-pi-1430zjgkdo4j`)

The `Controller` object also functions as a scheduler, where one behavior is active at a time. The active behavior is decided by the sensor data and its priority. Any old active behavior is suppressed when a behavior with a higher priority signals that it wants to run (Norris 2019).

```
import RPi.GPIO as GPIO import time
        class Behavior(self): global pwmL, pwmR
# use the BCM pin numbers GPIO.setmode(GPIO.BCM)
# set up the motor control pins GPIO.setup(18, GPIO.OUT) GPIO.
setup(19, GPIO.OUT)
pwmL = GPIO.PWM(18,20) # pin 18 is left wheel pwm pwmR = GPIO.
PWM(19,20) # pin 19 is right wheel pwm
# must 'start' the motors with 0 rotation speeds pwmL.
start(2.8)
pwmR.start(2.8)
```

There are two ways to use the `Controller` class. The first way is to let the class take care of the scheduler itself by calling the `start` method. The other way is to forcibly start the scheduler by calling the `step` method. (Norris 2019).

```
import threading class Controller():
def init (self): self.behaviors = []
self.wait_object = threading.Event() self.active_behavior_
index = None
self.running = True
#self.return_when_no_action = return_when_no_action
#self.callback = lambda x: 0

def add(self, behavior): self.behaviors.append(behavior)
def remove(self, index):
old_behavior = self.behaviors[index] del self.behaviors[index]
if self.active_behavior_index == index: # stop the old one if
the new one overrides it
```

```python
old_behavior.suppress() self.active_behavior_index = None
def update(self, behavior, index): old_behavior = self.
behaviors[index] self.behaviors[index] = behavior
if self.active_behavior_index == index: # stop the old one if
the new one overrides it
old_behavior.suppress()

def step(self):
behavior = self.find_next_active_behavior() if behavior is
not None:
self.behaviors[behavior].action() return True
return False

def find_next_active_behavior(self):
for priority, behavior in enumerate(self.behaviors): active =
behavior.takeControl()
if active == True: activeIndex = priority
turn activeIndex
def find_and_set_new_active_behavior(self):
new_behavior_priority = self.find_next_a havior() if self.
active_behavior_index is None or self.active_ behavior_index >
new_behavior_priority:
if self.active_behavior_index is not None: self.behaviors[self.
active_behavior_index].suppress()
self.active_behavior_index = new_behavior_priority

def start(self): # run the action methods self.running = True
self.find_and_set_new_active_behavior() # force it once thread
= threading.Thread(name="Continuous behavior checker",
target=self.continuously_find_ new_active_behavior, args=())
```

```
thread.daemon = True thread.start()
while self.running:
if self.active_behavior_index is not None: running_behavior =
self.active_behavior_index self.behaviors[running_behavior].
action()
if running_behavior == self.active_behavior_index: self.active_
behavior_index = None self.find_and_set_new_active_behavior()
self.running = False

def stop(self): self._running = False
self.behaviors[self.active_behavior_index].suppress()

def continuously_find_new_active_behavior(self): while self.
running:
self.find_and_set_new_active_behavior()
def_str_(self):
return str(self.behaviors)
```

Controller Category

The Controller category regularly occurs by means of allowing a large
range of behaviors to be applied to the usage of general-purpose methods.
The take Control() approach lets in a behavior to signal that it needs
to take control of the robot. The way it does this is discussed later. The
action() method is the way a behavior starts to control the robot. The
impediment-avoidance conduct kicks off its action() method if a sensor
detects an impediment in the robot's path. The suppress() approach
is used by using a higher-priority conduct to quit or suppress the
action() method of a lower-precedence behavior. This happens when
an impediment-avoidance conduct takes over from the normal forward
movement behavior by way of suppressing the ahead behavior's action()
method and using its own action.

Controller Type

The Controller type requires a listing or array of behavior objects that contain the robot's overall behavior. A Controller instance starts with the highest array index in the behavior array and examines the takeControl() method's return value. If true, it calls that behavior's action() method. If it is false, the Controller checks the next behavior object's takeControl() return value. Prioritization takes place by attaching index array values to each behavior object. The Controller category continuously rescans all the behavior objects and suppresses a lower-priority behavior if a greater-precedence behavior asserts the takeControl() method while the lower-priority action() method is activated. Figure 10-3 shows this procedure with all the behaviors that are eventually added.

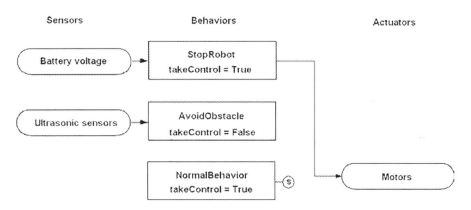

Figure 10-3. *Behavior state diagram (based on Norris 2019)*

It is now time to create a relatively simple behavior-based robot example.

Creating a Behavior-Based Robot

The `Controller` class allows a large range of behaviors to be applied via the use of the general-purpose methods. The `takeControl` method allows a behavior to signal that it wishes to take control of the robot. The `action` method is the way a behavior starts to manage the robot. The obstacle-avoidance behavior kicks off its `action` method if a sensor detects an impediment in the robot's path. The `suppress` method is used with the aid of a greater-precedence behavior to give up or suppress the `action` method of a lower-priority behavior. For instance, an impediment-avoidance behavior takes over from the normal forward-motion behavior by suppressing the forward behavior's `action` method and having its very own `action`.

It is now time to create a simple behavior-based robotic example. The `Controller` class requires a listing or array of `Behavior` objects that form the robot's basic behavior. A `Controller` instance begins with the highest array index in the `Behavior` array and tests the `takeControl` method's return value. If true, it calls that behavior's `action` method. If it is false, the `Controller` examines the next `Behavior` object's `takeControl()` method's return value. Prioritization occurs by looking at the index array values connected to each `Behavior` object. The `Controller` category continuously rescans all the `Behavior` objects and suppresses lower-priority behavior (Norris 2019).

Other Cognitive Architectures

Cognitive architectures in general refer to the following:

- An operational architecture that describes what the system does on a theoretical level.

- The system architecture can be developed by a manufacturer (or research group) and may look like a data-flow diagram.

- The technical architecture specifies the actual procedures and code structure.

In the following sections we'll look at some of the different kinds of architecture.

Reactive Cognitive Architecture

A reactive (or behavioral) cognitive architecture contains the following:

- Operational Architecture: describes *what* the system does at a high level, not how it does it

- Systems Architecture: describes how a system works in terms of *major subsystems*

- Technical Architecture: describes how a system works in terms of *implementation details*, language, algorithms, and code

Behavioral robotics, shown in Figure 10-4, is a basic "instinctive" architecture with sets of SENSE-ACT couplings called behaviors that get turned on/off based on stimulus; within this architecture there is no plan.

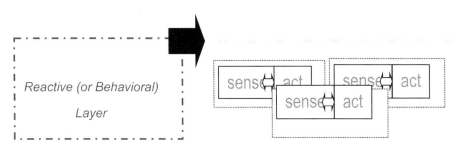

Figure 10-4. *Behavioral robotics (MIT Press 2019)*

Control theory is at a "lower" level; however, the behavior reacts very quickly because it is close to the sensor. This list is not exhaustive:

Reactive (Type 1, behaviors):

- Tightly coupled with sensing, consequently very quick

- Many nearly concurrent stimulus-response behaviors, linked together with simple scripting

- Action generated by sensed or external stimulus

- No situational awareness, no mission monitoring

- Models are of the vehicle, not of the environment

Thus far, behavioral organization suggests three layers of intelligence with distinctly different levels of perception, knowledge, planning horizons, and time scales. The AI robotics field has converged on PLAN, then SENSE-ACT, with LEARN being added later as needed at different points. Technically this is SENSE-PLAN, SENSE-ACT, but historically the sensing for planning, just like the execution monitoring, is lumped in with "PLAN." See Vemuru et al. (2019).

Canonical Operational Architecture

A canonical operational architecture is a potential method for how to program an intelligent robot. It's a type of model that aims to show data entities and relationships in the simplest feasible form in order to integrate processes across various systems and databases (Figure 10-5). Very often, the data exchanged across various systems rely on different languages, syntax, and protocols. The architecture consists of three layers that generally represent interaction, deliberative, and behavioral.

The interaction layer consists of procedural, functional, and ontological languages such as OWL. The deliberative layer consists of functional languages such as Lisp. The behavioral layer consists of procedural languages such as C, C++, and Java. Each layer has a different type of program structure.

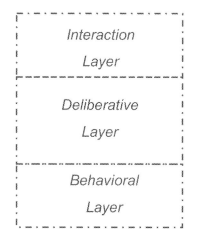

Figure 10-5. *Canonical operational architecture (Harbour et al. 2019; MIT Press 2019)*

How much intelligence the robot needs is determined by the following:

- What functions the robot needs to do, such as to generate, monitor, select, implement, execute behaviors, or to learn.

- What planning horizon the functions require; for example, Present, Present+Past, Present+Past+Future.

- How fast the algorithms need to update needs to be determined.

- Control theory may have to be implemented, such as using a closed world and guaranteed execution rates.

- Is the model for Local, or Global, or Both operations?

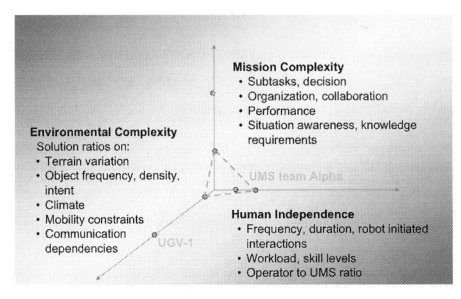

Figure 10-6. *Complexities (MIT Press 2019)*

System and Technical Architectures

A system architecture is developed by a researcher or study group and has three paradigms: hierarchical or reactive, or hybrid deliberative/reactive. System architectures can make the cognitive architecture for intelligence more tangible. They are very dependent on the implementation and involve sub-systems. The following are some additional features:

- Able to relate the functions in the canonical operational architecture to the five common sub-systems

- Classify a system's architecture as being either hierarchical, reactive, or hybrid deliberative/reactive based on 1) the relationship of the three AI robot primitives and 2) sensing handling

- Able to draw the hybrid deliberative/reactive system architecture

- Know three ways of generally organizing systems
- Understand these contributions to canonical system architecture

In order to make the architecture more tangible, again we add functions such as hybrid deliberative/reactive architecture, which yields three layers. The five sub-systems of the system architecture are as follows:

1. Navigation (generating)
2. Mapmaker, environment modeling
3. Planning (mission generation, executing)
4. Motor schemas (performing motor commands)
5. Perception, sensing, perceptual schemas

Technical architectures typically involve a new procedure, such as potential fields. The technical architecture is influenced by the operational and system architectures. The technical architecture contains specific algorithms, as well as control and knowledge structures. Sub-systems can further be reduced to two terms that are two attributes:

1. *Relationship*, which is how the three building blocks, or robot primitives, are arranged

2. *Content*, which is how sensing is handled (MIT Press 2019)

The Human Model

When one is deriving a cognitive model in order to achieve machine awareness, one must look to the human model. Let's start with the upper brain or cortex; here, we have reasoning over symbols and data about goals. Next, the middle brain converts sensor data into symbols data. Finally, the spinal cord and "lower" brain are where skills and responses occur. This can make a most abstract canonical operational architecture, creating a lot of interaction between the deliberative and behavioral (reactive) layers (Figure 10-7).

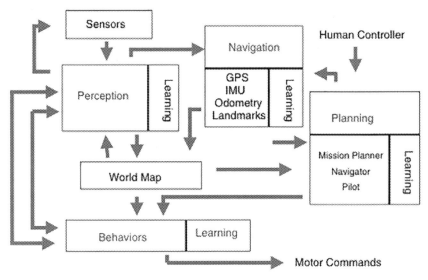

Autonomous behavior subsystems. Courtesy of Clint Kelley, SAIC.

Altered from Technology Development for Army Unmanned Ground
Vehicles 2002, National Research Council

Figure 10-7.   *Autonomous behavior*

Figure 10-8 shows the AI primitives (building blocks) with an agent.

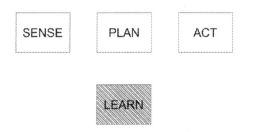

Figure 10-8.   *AI Primitives*

A Hierarchical Paradigm for Operational Architectures

Hierarchies are a natural way to organize, as they are not fundamentally rigid or inefficient (Figure 10-9). They are more flexible than centralized planning, and the goals and priorities are salient (Albus & Mystel 2001; Murphy, 2019).

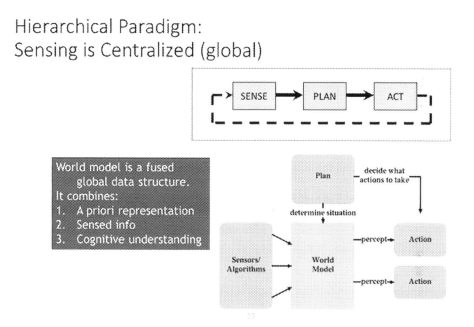

Figure 10-9. *Hierarchial paradigm (altered from Murphy 2019)*

Deliberative Architectures

This is where the AI robot receives and processes all of its available sensory information, then utilizes the appropriate internally stored knowledge it has, along with AI, and then reasons to create a plan of action. To do this, the robot must run a search through all potential plans until it finds one that will successfully do the job. This requires the AI robot to look forward and consider all feasible moves and outcomes. Unfortunately, this can be time consuming, creating latencies; however, advances in AI could shorten this time and allow the robot to act strategically.

Reactive Architectures

To allow the robot to respond very quickly to changing and unstructured environments, a reactive architecture can be utilized. This is where sensory inputs are tightly coupled to effector outputs, to allow for "stimulus-response" to occur. Animals in nature are largely reactive. However, this has limitations. Because it is stimulus-response only, this architecture has no working memory, no internal representations of the environment, and no ability to learn over time.

Hybrid Architectures

In a hybrid architecture, the best of both can be combined: reactive and deliberative. In it, one part/type/system of the robot's "mind/brain" plans, while another deals with immediate reaction, such as avoiding obstacles and staying on the road. The obvious challenges are effectively and efficiently bringing these two parts of the mind/brain together (Harbour et al. 2019). This may require an additional part of the AI robot mind/brain to be the "executive." (See Harbour et al. 2019.) Hybrid architectures can combine the responsiveness of reactive systems with the flexibility of purely deliberative systems, with the more traditional symbolic or deliberative methods yielding high levels of robustness. Solely reactive systems lack the ability to take into account a priori knowledge about the environment or to keep track of the past (what works and what does not work) via long-term memory. Nor do they use working memory; they just react. An example of a hybrid system could be a typical three-layer hybrid architecture: bottom layer is the reactive/behavior-based layer (Type 1 processing), in which sensors/actuators are closely coupled; the upper layer provides the deliberative component (e.g., planning, localization) (Type 2 processing); while the intermediate between the two could be called the executive layer or be contained in Type 2. (See Harbour et al. 2019.)

Summary

Task

There are four categories: Time, Subject of Action, Movement, and Dependency.

1. Time:

   - Fixed time. Ex.: Collect as many items in 20 minutes.

   - Minimum time. Ex.: Visit all buildings in area as fast as possible (minimize the time).

   - Unlimited time. Ex.: Patrol the buildings in the area.

   - Synchronization required. Ex.: Push two or more buttons at same time.

2. Subject of Action: This has two categories: object-based and robot-based.

3. Movement. Going-to or going-while, coverage or convergence.

4. Dependency: Independent, dependent, and interdependent.

References

Blank, D., Kumar, D., Meeden, L., & Yanco, H. (2003). Pyro: A python-based versatile programming environment for teaching robotics. *Journal on Educational Resources in Computing (JERIC)*, *3*(4), 1–es.

Clark, J.D., & Harbour, S.D. (2019). Unpublished.

Clark, J.D., Mitchell, W.D., Vemuru, K.V., & Harbour, S.D. (2019). Unpublished.

Chang, T.H., Hsu, C.S., Wang, C., & Yang, L.-K. (2008). On board measurement and warning module for measurement and irregular behavior. *IEEE Transactions on Intelligent Transportation Systems*, *9*(3), 501–513.

Dayan, P., & Abbott, L.F. (2001). *Theoretical neuroscience: computational and mathematical modeling of neural systems*. MIT Press.

Gerstner, W., & Kistler, W. (2002). *Spiking Neuron Models: Single Neurons, Populations, Plasticity*. Cambridge University Press.

Friston, K., & Buzsáki, G. (2016). The functional anatomy of time: what and when in the brain. *Trends in cognitive sciences*, *20*(7), 500–511.

Friston K. (2018). Am I Self-Conscious? (Or Does Self-Organization Entail Self-Consciousness?). *Frontiers in psychology*, *9*, 579. doi:10.3389/fpsyg.2018.00579

Harbour, S.D. (2022). SINCLAIR COLLEGE / UNIVERSITY OF DAYTON, AVT 4215 / ECE 595: Autonomous Systems in Aviation / Autonomous Systems and Artificial Intelligence. Lecture, University of Dayton, Ohio.

Harbour, S.D., & Christensen, J.C. (2015, May). A neuroergonomic quasi-experiment: Predictors of situation awareness. In: *Display Technologies and Applications for Defense, Security, and Avionics IX; and Head-and Helmet-Mounted Displays XX* (Vol. 9470, p. 94700G). SPIE.

Harbour, S.D., Rogers, S.K., Christensen, J.C., & Szathmary, K.J. (2015, 2019). Theory: Solutions toward autonomy and the connection to situation awareness. Presentation at the 4th Annual Ohio UAS Conference. Convention Center, Dayton, Ohio. USAF

Harbour, S.D., Clark, J.D., Mitchell, W.D., & Vemuru, K.V. (2019). Machine Awareness. *20th International Symposium on Aviation Psychology*, 480–485. https://corescholar.libraries.wright.edu/isap_2019/81

Kidd, C., & Hayden, B.Y. (2015). The Psychology and Neuroscience of Curiosity. *Neuron*, *88*(3), 449–460.

Kistan, T., Gardi, A., & Sabatini, R. (2018). Machine learning and cognitive ergonomics in air traffic management: Recent developments and considerations for certification. *Aerospace*, *5*(4), Article Number 103.

Loewenstein, G. (1994). The Psychology of Curiosity: A Review and Reinterpretation. *Psychological Bulletin*, *116*(1): 75–98.

Mitchell, W.D. (February, 2019). Private communication.

Murphy, R.R. (2019). *Introduction to AI robotics*. MIT press.

Norris, D.J. (2019). *Beginning artificial intelligence with the Raspberry Pi*. Apress.

Li, L.S., Hansman, R.J., Palacios, R., & Welsch, R. (2016). Anomaly detection via Gaussian mixture model for flight operation and safety monitoring, *Transportation Technologies, Part C: Emerging Technologies*, *64*, 45–57.

Pour, A.G., Taheri, A., Alemi, M., & Meghdari, A. (2018). Human-Robot facial expression reciprocal interaction platform: Case studies on children with autism, *International Journal of Social Robotics*, *10*(2), 179–198.

Rogers, S. (2019). Unpublished.

Sharpee, T.O., Calhoun, A.J., & Chalasani, S.H. (2014). Information theory of adaptation in neurons, behavior, and mood. *Current opinion in neurobiology*, *25*, 47–53.

Vemuru, K.V., Harbour, S.D., & Clark, J.D. (2019). Reinforcement Learning in Aviation, Either Unmanned or Manned, with an Injection of AI. *20th International Symposium on Aviation Psychology*, 492–497. https://corescholar.libraries.wright.edu/isap_2019/83

Xu, S.T., Tan, W.Q., Efremov, A.V., Sun, L.G., & Qu, X. (2017). Review of control models for human pilot behavior. *Annual Review in Control*, *44*, 274–291.

Zhao, W.Z., He, F., Li, L.S., & Xiao, G. (2018). An adaptive online learning model for flight data cluster analysis. In: *Proc. of 2018 IEEE/AIAA 37th Digital Avionics Systems Conference, IEEE-AIAA Avionics Systems Conference* (pp.1–7). London, UK.

CHAPTER 11

Geospatial Guidance for AI Rover

Throughout this chapter, you will learn about geospatial data, how Geographic Information Systems (GIS) is incorporated into AI, and the rover's navigation awareness. You will come to understand and develop the first navigational GIS routines for this rover.

Where you are located can be a simple question to answer for humans. We gather data from the world around us using our senses. The brain processes this information into a location. We then use additional data to confirm that location. The AI rover will need to do the same thing using systems and programming loaded onto the rover.

The Need for Geospatial Guidance

The AI rover will use geospatial data to locate itself. This data is not unique; it is used everywhere and in many different fields, such as transportation, archeology, earth sciences, agriculture, and many more. Geospatial data is anything that has a spatial component related to a location. This location could be on Earth, in the oceans (beneath or on the surface), in the solar system, on the moon, or other celestial bodies, and the hot-topic area that

D. A. Blubaugh et al., *Intelligent Autonomous Drones with Cognitive Deep Learning*, https://doi.org/10.1007/978-1-4842-6803-2_11

a lot of geospatial work is being performed in relation to is Mars. The many rovers and UAS on Mars are all using geospatial data to operate in that environment.

To understand geospatial data, we have to have an understanding of geomatics. The umbrella discipline of geomatics provides the foundational tools that geospatial data uses. These tools include storage, analysis, collection methodologies, and dissemination techniques. Geomatics involves several sub-disciplines. Each of these disciplines is sophisticated enough to be its own specialty in and of itself. Many professionals will focus their entire careers on just one of these disciplines. Geomatics contains the following disciplines: surveying, geodesy, cartography, photogrammetry, remote sensing, and geographical information science (GIS). There are many more that will not be discussed in this chapter. The AI rover will use many facets of each of these disciplines. Thus, an understanding of each of them is warranted.

The first discipline and the foundation that all others use is geodesy. Geodesy is the understanding of the accuracy of our measurements, understanding the Earth's size orientation in space, and knowing how different phenomena impact our location. Under geodesy, datums are created for the environment. These datums are used as a starting point or a location where zero is measured. There are many different types of datums. They can be horizontal or vertical, tidal, local, state, or global. These datums can be based on the geode or ellipsoid and can be used in conjunction with one another. Geodesy also includes coordinate systems. Common coordinate systems are latitude and longitude, Universal Transverse Mercator (UTM), and Military Grid Reference System (MGRS). These coordinate systems allow the rover to calculate its location and for that location to be easily found by others. A map is then used to depict this locational data. Representing a globe poses some problems with two-dimensional maps; specifically, the distortions created by defining a curved surface on a two-dimensional map. Only a globe would be able to preserve all the qualities and minimize distortions. This issue can be solved by using projections. These projections are cylindrical, conical, and azimuthal. You now know some of the critical principles behind geodesy.

The next discipline is cartography. Cartography is the art, science, and technology of creating maps using information from geodesy and other disciplines. Cartography in the modern age uses computers to create maps, and on these we can display many different types of data. These maps are designed with the end user in mind. Cartography products can be topographic maps with contour lines, shaded relief, planimetric, and many more. No matter what product is generated, how the end user will use the product is the primary focus of cartography. Would a pilot need to know the distance between sewer grates?

Another discipline under geomatics is surveying. Surveying is the direct measurements taken using systems such as photogrammetric ground control, GPS, and traditional field surveying methods. Traditional field serving techniques for horizontal control use traversing, triangulation, and trilateration. In the vertical datum, differential and trigonometric leveling are used. Already you can see the complex nature of geomatics, where one discipline's information is required for the others.

GPS is another system used by geomatics. GPS has simplified data collection and displays a location on the Earth's surface. It uses information from GIS, surveying, and geodesy in order to display this position. Humans, in general, have come to rely on GPS and have a basic understanding of GPS. GPS is a constellation of satellites that send a signal down to Earth, which is received by a GPS receiver, which then displays our location on a moving map. However, there is much more to GPS than what the end user sees. Typical systems today use a combination of different GPS systems. This is referred to as the Global Navigation Satellite System, or GNSS. The GNSS can use the USA-based GPS, but it also uses other countries' GPS systems. The GNSS is preferred because of its higher accuracy.

The more satellites in view of the receiver, the better accuracy you will get in the form of an average position. For any GPS, there are three main components. The first is the control segment. The control segment monitors the health of the satellite constellation and sends corrections when warranted. The spaceborne is the next segment. These are the GPS satellites themselves. GPS satellites are placed in orbits that allow the entire Earth to be covered at any one time. These orbits ensure that a minimum number of satellites are in view to localize your position. The last segment is the user. This segment requires a receiver that can receive the GPS satellites' specific signals. Computational processing and a display are also a part of this system. The primary function of GPS is triangulation and uses the time-of-arrival concept. The satellite transmits a signal that travels from the satellite to the receiver. The receiver will process the distance from that specific satellite. The GPS uses this distance and distances from additional satellites, and the receiver unit can then triangulate your position on Earth. You need three satellites in view for a horizontal position, and to add a vertical position, the number increases to four satellites.

The last component of geomatics this chapter will discuss is remote sensing. Remote sensing relies on gathering data without being in direct contact with the analyzed object. This can be accomplished with sensors that use areas in the visual spectrum as well as outside it. There are a variety of sensors that could be used for remote sensing, including electro-optical, infrared, multispectral, hyperspectral, chemical, biological, or nuclear. These are all passive sensors, meaning they receive information but do not transmit. An active sensor, such as ultrasonic, LiDAR, and RADAR, transmits energy that bounces off an object and is then received. These sensors, if attached to the AI rover, provide two benefits. The first is that these sensors could be coupled with AI sensing to identify and avoid objects that could stop the rover in its path. The second benefit is that all the data collected could be analyzed after the rover returned to the entrance.

The multispectral and hyperspectral sensors gather data in targeted bands of the EM spectrum and produce a spectral signature. This spectral signature can be compared to known spectral signatures to identify what the object is made out of or composed of, which would be highly beneficial when classifying objects/obstacles. RADAR, LiDAR, and ultrasonic sensors are all primary sensors that can map out obstacles.

The AI rover will use a variety of sensors and add-ons to navigate and map the catacombs of Egypt. This function, at its core, requires the use of many of the principles of geomatics. Thus, an understanding is needed of these core functions.

The significant difference the AI rover will encounter is the known versus the unknown. Programing a rover to navigate a known space on the surface is relatively easy. Let the rover acquire its starting GPS coordinates, and the operator or AI will plot the mission in the form of waypoints. The autopilot and connected mission planning software make all the calculations necessary. The AI rover starts the mission and monitors and makes corrections as the mission progresses. The added benefit is that a human operator can monitor the AI rover's progress via a data-link and override or adjust as necessary. Ground, air, or sea, this is essentially what uncrewed operators do to gather data. In an open space, with access to the operator and GPS, this operation is ideal and has led to a more efficient gathering of geospatial data.

Some different approaches can be used to gather data from the unexplored catacomb in Egypt. A simple rover could be built that operates like first-generation robotic vacuum cleaners. These vacuums would bounce into an object, turn a set number of degrees, and continue vacuuming the house. This created a random pattern of cleaning an uncleaned area. Large sections of rooms could be missed if the robot wandered into an adjacent room. This form of operation is inefficient. In the catacombs, this robot would not be able to return to the starting point. Thus, all of the collected data would still be in the catacomb and of no use to researchers.

The better approach is to use current terrestrial rovers. This rover would have the ability to use GPS to locate itself at a starting point. Couple that with real-time kinematics (RTK), and the rover would be able to locate itself to mere centimeters. An RTK system uses a fixed base station and a roving station. Both of these receive GPS or GNSS signals, and the base station calculates a correction that is sent to the roving RTK station to update its position. The issue here is that we are operating in a cave, where signal degradation or blocking would not permit the corrections to be received by the rover. In addition, being underground, the roving RTK station would not receive a GPS or GNSS signal.

The last approach that could be utilized is a tethered rover. This system allows for an operator to manually pilot the rover. This system will still requires the use of onboard sensors to spot objects. The main issue here is that the tether would need to be very long and could get caught on obstacles. Having a long tether could lead to the loss of the system inside the catacombs. There are benefits of this system, such as that the tether could be powered, providing indefinite operation of the rover.

Needless to say, none of these approaches will work very well. Thus, the need for a fully autonomous, AI-driven rover that incorporates geospatial guidance and data logging into its operation.

Why Does the AI Rover Need to Know Where It Is?

Our AI rover is operating in an *almost* entirely unknown environment. There is one location that the rover will know when it is powered up. That is the starting location. This starting location has many different names associated with it. Start point, null point, return home point, home location, etc. Having a known location to start from assists the rover in applying the datum to the overall operation. A datum is a frame of reference for measurements that all the systems onboard the rover will use to navigate, gather, and export data collected. A datum includes the coordinate system and projection used in the form of metadata.

Once the AI rover leaves the starting point, the unknown is encountered. The different tunnels, cave-in locations, large rocks, human-made objects, even the composition of the ground the rover is traversing is unknown. Humans programming the rover will be able to make assumptions based on other catacombs in the area. However, these are just assumptions. The rover will need to be able to sense the world around it. These sensors come in many forms, such as ultrasonic ranging, light detection and ranging (LiDAR), and stereoscopic vision systems.

Ultrasonic sensors are a form of echolocation, similar to what bats and whales use to navigate. These sensors send out sound waves that humans cannot hear. This wave strikes an object, reflecting some of that energy back to the source. This reflection is then what is used to calculate the distance from the object.

LiDAR and RADAR are similar means of active remote sensing. The difference between these sensors is what portion of the electromagnetic spectrum these sensors actively transmit. LiDAR uses light, and RADAR uses radio waves. Both of these use the time-of-flight concept to calculate the distance to the obstacle. Many passes of these returns can be used to create a point cloud. The drawback to these sensors is that they generate a significant amount of data. This will require additional data transfer and storage for further analysis. However, if just used for navigation, there is no need for extra storage.

Stereoscopic vision systems are the robotic version of human eyes. Humans have taught computers to "see." These sensors incorporate two electro-optical sensors that are a known distance apart. Then, they use the parallax-error concept to calculate the distance between the rover and the obstacle. This principle is precisely how humans process depth perception.

These sensors will sense obstacles and generate raw geospatial data. Geospatial data combines two pieces of information: the object's location as it applies to the datum, and an attribute of the object. The geospatial data is then sent to the AI to make a decision in regard to the object. If it is a rock, then which way should the rover turn to avoid it? Suppose it is two different tunnels; which tunnel will it continue down? These are immediate decisions that the rover will need to make. However, the AI accepting this raw data and making decisions is just one part of the system's knowing where it currently is. The AI rover will also need to store the raw geospatial data and create a detailed map of the locations of obstacles. This data is then used to make the most efficient return to the home route. The added benefit is that this map and additional data can then be downloaded from the rover for archeologists to use in their study of the catacombs.

How Does GIS Help Our Land-based Rover?

At its core, a geographical information system (GIS) is software and hardware that takes data in and allows it to be processed, visualized, analyzed, and published. The GIS software that is loaded into the AI rover will take the raw geospatial data from the sensors and Robotic Operating System (ROS). This data is then processed into useful information that the AI can use for navigation and the creation of maps of the catacombs that researchers can use. GIS allows for the creation of maps and the ability to analyze or query the data to answer questions we might have. Heights and distances are some of the information that can be gathered via GIS. GIS is cyclical in nature. The more data we can feed the software, the better the products and answers it will produce. Thus, the rover will constantly be providing information to the software, and it, in turn, will create maps that the AI rover can use for navigation.

Which GIS Software Package Do We Use, and Can It Be Used with an ROS-based Rover?

Many GIS software packages can be utilized with the AI rover. You may have heard of some of them, such as Google Earth or ArcGIS. The software that you will load into the rover is QGIS. QGIS is open-source GIS software that has been around since 2002. This software was chosen due to its easy integration into ROS to access live data generated by the sensors. This is a vital function, as the sensors are going to be feeding QGIS live data. QGIS is plugin based and can run on a variety of different platforms, such as Linux, Ubuntu, Windows, and MacOS.

Can GIS Be Embedded Within Our AI-Enabled Rover?

The short answer to this question is yes, it most certainly can! There are quite a few options available or custom-coded to work with the rover. However, simplicity and cost are key here. Using open-sourced products like QGIS allows one to embed GIS into the rover. As QGIS has many plug-in options, we will employ one of these in the AI rover. It is called QGIS-ROS. This simple plug-in connects ROS to QGIS and requires no coding skills. Simply load the plug-in into ROS. Once the plug-in is loaded, then ROS has access to powerful GIS software.

Summary

We live in a spatial world, with objects having a geographical location that can be mapped, analyzed, processed, and published. However, a common frame of reference is needed. This is the fundamental principle behind geomatics. Using geomatics' tools will allow our rover to operate in the unknown. The AI rover needs to track its location and gather data that can be used for further analysis in GIS software. The use of open-source products like QGIS means we can complete the mission cost-effectively. Having the AI rover's ROS at its core integrated via the QGIS plug-in means the data is already formatted and can be easily analyzed after the mission has been completed.

EXTRA CREDIT

1. Watch the following video regarding QGIS integration in ROS: https://vimeo.com/293539252

2. QGIS is open source; download it and use the tutorials and videos to understand the interface and the operation of the software.

    a. http://www.qgis.org

    b. Manual: https://docs.qgis.org/3.10/en/docs/ training_manual/index.html

    c. Tutorials/training: https://training.datapolitan. com/qgis-training/Introduction_to_GIS_ Fundamentals/#1

3. Download the QGIS-ROS plug-in from GitHub: https://www.Github.com/locusrobitics/qgis_ros

4. If you have found GIS exciting and would like to learn more, contact your local community college, college, or online college and take an introductory class. There are many options for careers that one could specialize in.

CHAPTER 12

Noetic ROS Further Examined and Explained

We will now further explore the realms of the Robot Operating System (ROS) ecosystem to closely examine the capabilities available in Noetic ROS. This chapter covers ROS features, utilities, facilities, and tools more deeply and supplements Chapters 3 and 4.

Objectives

The following objectives will be covered about ROS:

- Understand the rationale, philosophy, and purpose of ROS.

- Explore ROS facilities, such as graphs, `roscore`, `catkin_make`, packages, `rosrun`, `roslaunch`, and workspaces.

- Explore ROS topics, services, and actions.

© David Allen Blubaugh, Steven D. Harbour, Benjamin Sears, Michael J. Findler 2022
D. A. Blubaugh et al., *Intelligent Autonomous Drones with Cognitive Deep Learning*,
https://doi.org/10.1007/978-1-4842-6803-2_12

- Compare and contrast ROS1 and ROS2 for the Raspberry Pi 4.

- Debug faulty ROS programs with `rqt_console`.

ROS Philosophy

The fundamental idea behind Noetic ROS is to allow people to create robotic applications without reinventing the wheel. This capability involves sharing codes, algorithms, libraries, and ideas to efficiently create more complex robotic applications. There are well over 2,000 hardware and software applications for ROS. These include drivers to receive data from sensors such as LiDAR and to send commands to the rover's actuators. We also have an expanding collection of algorithms that allow SLAM, navigation, image-processing, object-detection, path-finding, and neural cognitive solutions with ROS. Noetic ROS distributes the computational requirements and divides the workload across multiple ROS programs or nodes. However, we must be careful with memory constraints and keep the total number of nodes to no more than 7. ROS uses multiple debugging tools to correct distributed workflows following any discovered faulty behavior. These tools manage the complexity of running multiple ROS node programs. These advantages of ROS allow ROS developers to save money, research effort, and, above all, time.

Like the other ROS environments (Kinetic, Indigo, etc.), the Noetic ROS environment allows one to use the UNIX philosophy of software development. The UNIX philosophy aims to create stable, modular, and reusable software. Another way to state this is that it helps to build small programs that do one thing, and do it well. ROS does this with ROS node programs, where each node program does one thing, such as SLAM or object detection, and does it exceptionally well. The following paragraphs describe the many ways in which ROS adapts and exploits the UNIX philosophy for software development.

- ROS systems consist of small computer programs called nodes that interact with each other and exchange messages. These data and command messages establish connections between each of the program nodes. There is no need for a central routing service to manage these messages, which allows our AI rover ROS application to scale. We can incrementally do so by adding SLAM, perception, object vision detection, and sense and avoidance as individual nodes. Many development environments such as Python Pycharm and GNU toolsets support ROS node development.

- ROS incorporates both open source and proprietary software to allow for even greater robotic capabilities and functionalities. Noetic ROS also encourages developers to use foreign libraries, such as TensorFlow, that can be used to both send and receive messages to other ROS node programs. We can then take the same library offline and test it outside an ROS node to determine if it is working correctly. Noetic ROS also supports both open source and proprietary software libraries. These two types of software can operate with each other by running inside ROS nodes.

ROS Fundamentals

For this section, we will quickly review the underlying architecture of Noetic ROS. ROS allows for multiple different programs to communicate with each other via messages. We will now look even closer at the underlying architecture that allows this to happen. We will also see how we can incorporate well-tested software for ROS. The reuse of software will allow us to address the many challenges the AI rover might face.

One of the original problems that motivated the creation of ROS was the "fetch an item" challenge. This challenge is very similar to the original mission requirements of the AI rover to explore the Egyptian catacombs to find and identify artifacts within those catacombs. The main difference is that the AI rover will not have a manipulator's arm. However, it will have a camera, laser/ LiDAR sensor, gyroscope, and inertial measurement unit (IMU). We could easily add a manipulator arm to the AI rover, adding additional complexity to this textbook. However, we will provide a supporting website for this textbook that will review end-of-arm tooling and manipulator's arms to safely extract objects found by the AI rover in its environment. The task of the rover is to navigate an environment, find an item, identify an item, and locate an item's position. The rover then must return to the initial starting point once it locates all items. This fundamental task of the AI rover will highlight many particular and peculiar requirements for the software application controlling the AI rover. Some of the more detailed requirements of the robotic software development for the AI rover are as follows:

- A Noetic ROS application must be decomposed into separate sub-systems such as navigation, vision, sense and avoid, path-finding, decision analysis, and decision execution.

- We could also recycle these very same sub-systems to use them in other robotic applications, such as military patrols and security.

- We should be able to operate the identical sub-systems on different robotic hardware platforms with little or no alterations to the underlying source code.

The first component of ROS that we should review is the graph of a ROS node, which will allow us to describe concurrently running programs that are sending and receiving messages to one another. We will be using graphs to describe the interconnections between the ROS programs, described as nodes. The graph allows us to manage the complexity of the ROS application. The graphical lines defined between these nodes are also where messages will transfer data and commands between nodes. Therefore, by using a graph of the ROS application, we can instantly see the complexity and features of the entire application. We can also understand the functionalities and what nodes are transferring what type of data or commands to the other nodes. We can then see that each ROS program is a specific node and that it is just one piece of a complicated robotic system.

An ROS graph node represents a self-sustaining Python program that sends or receives data or command messages. The same messages connecting these nodes represent a graph edge. ROS nodes are typically Portable Operating System Interface (POSIX) processes, and the edges are TCP ethernet connections with IP addresses. The separation of the Python source code controlling the LiDAR, camera, IMU, gyroscope, and neural processing into separate ROS nodes allows the ROS application to be far more resistant to failure. Therefore, if a single or multiple ROS nodes were to fail, these failures would not cause the entire ROS application to fail. However, we need to examine one potential "Achilles' heel" of any ROS application, and that would be the roscore. The roscore is the primary ROS master node program that allows each ROS node to find one another and forms a kind of operating system kernel for ROS.

The roscore master node is similar to any OS kernel as it provides information services for the ROS nodes to transmit and receive data and command messages to and from other ROS nodes. The roscore node is critical for any ROS application since this master node allows the other nodes to connect correctly. The roscore master node, during the powering and initialization phase, accepts and documents all other ROS nodes' required message streams. The roscore node also adds new ROS nodes to the main ROS application. The roscore node will form the correct message stream between the newly added ROS node and the other needed nodes for connection.

> **Note**   However, we need to be clear that once the `roscore` master node has established the connections between the nodes, the `roscore` node does not directly handle the exchange of data and commands between the nodes. This prevents data-block issues within the ROS application (AI rover). The issue of data block could only occur if the `roscore` node itself had to handle the exchange of messages completely. Instead, the ROS nodes handle data exchange and command messages between them. The `roscore` interaction only occurs once the nodes have established contact.

The `roscore` master node program is the fundamental foundation for *all* ROS applications (rover). This foundation allows nodes to find other nodes in an ROS application. The `roscore` master node also provides a parameter server, which allows the configuration of ROS nodes. The parameter server allows nodes to store and retrieve arbitrary data structures, such as the learning rate parameters for neural networks. The following vital foundations to consider for any ROS application (rover) would be creating the workspace for the ROS application source code to reside, the organization of ROS packages, and how the ROS build system with `catkin` operates.

Noetic ROS Catkin

The ROS build system, or the `catkin` system, is a set of tools that ROS uses to generate executable programs and libraries for our ROS application (AI rover). The libraries that we will use include TensorFlow or Keras libraries. These libraries develop deep neural network routines for the AI rover. Since we are using primarily only Python as the development language for the AI rover, we will not need to know the compilation process for `catkin` as we would if we were using a compiled programming language such as C/C++. However, for additional information regarding the catkin ROS build in a C/C++ system, please refer to http://wiki.ros.org/catkin?distro-noetic.

The catkin ROS build environment uses a suite of toolsets, CMake macros, and Python programs to enhance the Unix CMake workflow to create unique ROS applications. CMake produces two crucial files, CMakeLists.txt and package.xml, with specific information needed for ROS to work correctly. We will quickly review the tools needed to modify and append information to these two files for a successful ROS creation.

Noetic ROS Workspace

We must now further test the catkin workspace catkin_ws created in Chapter 3. We can create multiple workspaces. For example, in Chapter 6, we were able to partition the Python and launch source code into separate controller and simulation workspaces. However, we can only develop and test source code within a single workspace at any one time.

We must check again to determine if our system-wide ROS setup script STARTrosconfig.sh file is still working from Chapter 3. We need to consistently execute this script before making or compiling any workspace directory, as shown in Figure 12-1. For more information about creating the setup script, please refer to Chapter 3.

Figure 12-1. *STARTrosconfig.sh from Chapter 3*

We must be sure that we can also develop and initialize the catkin workspace, seen in the orange box. We then enter the commands seen in Figure 12-2.

423

Figure 12-2. *Initializing the catkin workspace*

The catkin_ws initializes once both the workspace directory and the source code (/src) directory are created. Again, this source code directory is where the ROS application (rover) source code will reside. The catkin_ws/src directory then requires the catkin_init_workspace command. This ROS command will generate the CMakeLists.txt file in the /src directory, as shown in Figure 12-2. The generated CMakeLists.txt file also contains a set of directives and instructions describing the ROS project's (rover) source files and targets (executables, libraries, or both). When you create a new project or workspace, the catkin_init_workspace command automatically generates the CMakeLists.txt file and places it in the project root directory (~/catkin_ws/src). You can point the Gedit editor to the CMakeLists.txt file and choose "Open as Project."

We will generate the (/build) and (/devel) directories to allow the catkin system to save libraries and executable programs developed with the C/C++ programming language. However, since we will be using the Python programming language, we will only be using the (/build) directory occasionally. To generate the (/build) and (/devel) directories, we must now enter the commands found in the orange box in Figure 12-3.

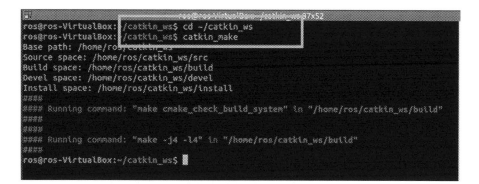

Figure 12-3. *Catkin_make command*

Noetic ROS Packages

The organization of the ROS software for the rover is done in packages. All Noetic ROS packages will reside in a workspace's /src directory, including a CMakeList.txt file and a packages.xml file. Both files discuss how catkin_ws should interact with any ROS package to create the foundation for ROS applications. These same packages also contain code, data, and documentation. To create an ROS package, one would need to use the following commands:

```
cd  ~/catkin_ws/src
catkin_create_pkg ai_rover rospy
```

All packages must reside in the /src directory to keep source code files efficiently organized. We have already invoked the catkin_create_pkg to create the ai_rover package. However, every ROS package is also highly dependent on ROS package dependencies, such as rospy. There are many other examples of ROS dependencies, such as NumPy, TensorFlow, and OpenCV. Also, if the main ROS dependency (rospy) requires other libraries as additional dependencies (OpenCV), you can list them in the following manner in the command terminal:

425

```
cd  ~/catkin_ws/src
catkin_create_pkg ai_rover rospy opencv
```

The `catkin_create_pkg` command creates a directory with the same name (`ai_rover`), `package.xml` and `CMakesLists.txt` files, and a `/src` directory for our future AI rover Python script files. Later in this chapter, we will go over the metadata (or data concerning data) found in the `packages.xml` and `CMakesList.txt` files.

Now that we have created a ROS package, we can place the Python nodes (neural networks, vision processing, and sense-and-avoid capabilities) into the ROS packages' `/src` directory. We will also conveniently develop future launch, world, and robot URDF description files to be organized into their own unique and separate directories. Now that we have a `/src` directory for our Python script nodes to reside in, we must review the terminal ROS commands necessary to execute these Python nodes to control our AI rover.

Noetic ROS rosrun

The entire infrastructure of ROS is dependent on the creation of the packages that we just reviewed in the previous section. A ROS package is a collection of Python source code, launch files, world models, and robotic description files built and distributed together. This allows us to develop concurrent processing systems for ROS, since each package could hold multiple different compiled Python script routines that could be executed in parallel. Each compiled Python script file goes into an ROS node. Therefore, we could start and manage any number of different ROS nodes simultaneously to create the intelligence of our AI rover.

If we were to revisit Chapter 3, we would see the example of the "talker" and "listener" ROS testing nodes, first used in Chapter 3 to test whether the installation of Noetic ROS was successful. Both of these ROS nodes were attached to the `rospy_tutorials` package. The one issue with packages is finding or locating their directories. We use the ROS command utility `rosrun` to find any needed directories.

Note   For more information regarding the "talker" and "listener" examples, please refer to Chapter 3.

Once we have initiated the "talker" node to send messages to the "listener" node, we can use the ROS query terminal command-line tools to determine the current status of the ROS system (AI rover). One of the primary and most potent ROS query command-line tools is `rostopic`. The `rostopic` command prints out the current list of active topics to the console, which is critical information for determining if our ROS system is working. If we were to leave the three terminal windows of the `roscore`, the "talker," and the "listener" nodes open and running, we could then open a fourth terminal window and enter the following command of `rqt_graph`. Once we enter this command, we should see the following display (Figure 12-4).

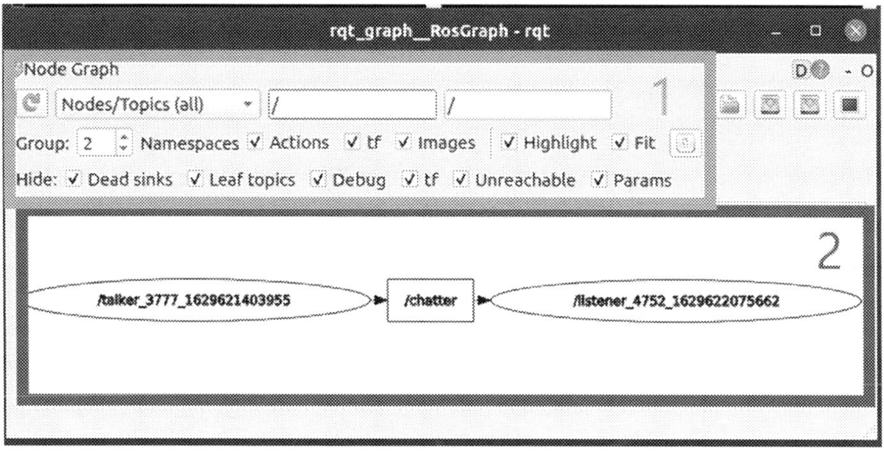

Figure 12-4. RQT GRAPH tool diagram

Once you see the display in Figure 12-4, it would be wise to first click on the auto-regenerate icon in the top-left corner of the orange-colored box. We do this to have the most recent version of any messages passed in our current ROS system. We also click on the auto-regenerate icon every time a ROS node is removed or added to the ROS application. The auto-refresh icon is shown in Figure 12-5. Once we click on the auto- regenerate icon, we should see that the graph in the blue-colored box is the latest and currently active ROS application. These tools will help us develop, test, and deploy ROS1 and future ROS2 applications onto the Raspberry Pi 4. The Raspberry Pi 4 will serve as the "brains" of our brave rover.

Figure 12-5. *Auto-regenerate icon*

Building the Rover's Brains

We have come to the end of this textbook. We must now install the Raspbian operating system or Ubuntu 20.04 LTS for the Raspberry Pi 4. These operating systems support ROS1 Noetic. In the case of ROS2 or Foxy, we only install the 20.04 Ubuntu server. As part of installing ROS1 or ROS2, we must be able to connect to the Raspberry Pi 4 through our development laptop or desktop computer to control and monitor the rover as it conducts its mission with a pre-loaded Raspberry Pi 4 onboard. But first, we must review how ROS1 and ROS2 differ to know which ROS version to load for a specific rover mission. ROS1 Noetic might be better if the rover mission is not mission critical. ROS2 Foxy might be required if the software development is lengthy or complicated (involving swarm intelligence between multiple rovers) and requires object-oriented programming.

ROS1 Versus ROS2

You may be asking how ROS1 and ROS2 differ? What is ROS2, and what advantages does it provide our intrepid rover?

This section will give you a practical overview of what has changed and what is new. We will focus on the main differences between ROS1 and ROS2. We'll try to be brief.

We will only use the term *ROS* when talking about ROS—aka Robot Operating System—in general: ecosystem, community, etc. However, we will use *ROS1* and *ROS2* when discussing the specific ROS versions for this chapter section.

Since 2007, the ROS1 development team has used their years of experience and lessons learned to determine what features to add and improvements to make. Unfortunately, adding all those modifications into ROS1 would have caused instabilities in ROS1 Noetic. So, ROS2 was developed from scratch and is a completely new ROS. This new variant of ROS incorporates the lessons learned with ROS1. ROS1 areas for improvement include safety and security. ROS needs to address these areas before ROS enters industrial applications.

Rover ROS1 or ROS2?

There are three main areas that separate ROS1 Noetic and ROS2 Foxy; let's closely review them. We should first consider that ROS1 Noetic for the rover is the last ROS1 variant, and support will end in May 2025. The ROS Noetic version provides support for the Python3 language. Noetic ROS has a substantial base of developed libraries for sensors such as LiDAR that will take some time to be ported to ROS2 Foxy. Meanwhile, ROS2 Foxy support will only last for a year since a new version of ROS2 will be available each year. ROS2 is now stable, so we can develop the rover missions. The first difference between ROS1 Noetic and ROS2 Foxy involves the ROS nodes.

ROS1 Nodelets or ROS2 Components

For simplicity, we will not review the use of C/C++ programming development for ROS nodes. We will only concentrate on the details of the Python language for this and other Chapter 12 sections. For further details on the C++ programming of ROS applications, please refer to the ROS.org website.

Consider using ROS callback functions in class nodes are used as the foundation of ROS1 nodes. As such, the development of ROS1 nodes is unstructured. However, in ROS2, structured development is made possible using a class template for the node object. ROS2 inherits its functionalities from the base class of the node. ROS2, using a good object-oriented programming (OOP) and modular approach to node creation, will help the creators of these nodes with potentially complicated node interactions. ROS2 is far more compatible with OOP development. ROS2 also allows you to convert groups of nodes into components. Node components can also be represented as UML diagrams, which further enhances the modular design in ROS.

A ROS1 node is also an executable. There are also ROS1 nodelets, which launch multiple nodes in the same executable with a single launch file—perfect for limited hardware resources or sending messages between nodes. In ROS2, these software entities are now called components. A component is simply a slightly modified node class from OOP. With ROS2, you can now handle multiple nodes from the same executable using components.

ROS2 components allow us to create efficient ROS2 applications. Python scripts allow for starting components from a launch file, the terminal, or another executable. There also exists programmable intra-process communication to remove any required ROS2 communication overhead.

ROS1 and ROS2 LaunchFiles

ROS1 and ROS2 LaunchFiles allow you to start all your nodes from one file. You can start a standard node, a nodelet, or a component—LaunchFiles, developed in ROS1, are programmed with XML.

In ROS2, Python is now used to write LaunchFiles. Python allows you to start one or more nodes. Python also allows LaunchFiles to be more adaptable to changing robotics needs, such as changing sensors to the rover or even replacing the rover design with an alternative design.

ROS2 LaunchFiles can be programmed with both XML and Python if you want to. But using Python allows more modularity and has become the standard ROS2 procedure for developing and initiating LaunchFiles. The following links describe how one develops LaunchFiles for ROS1 (`http://wiki.ros.org/roslaunch`) and ROS2 (`https://roboticsbackend.com/ros2-launch-file-example/`).

ROS1 and ROS2 Communications

The first significant difference between ROS1 and ROS2 is that there is no ROS master node in ROS2. In the past, ROS1 required starting the ROS master node first before any ROS application started. The ROS1 master node also acted as a controller for the message topics between ROS1 nodes. However, in ROS2, there is no longer any ROS master node. This lack of a master node allows for actual asynchronous distributed behavior in that each node can interact with other nodes. Each node in ROS2 is independent and not dependent on a global master.

ROS1 and ROS2 Services

In ROS1, services are only synchronous. That means that the ROS master node must respond to the responding ROS node promptly, or fail to do so. However, in ROS2, all services are asynchronous. You can add a callback function that will only respond once the responding node or service responds when you call a service. This feature prevents the resource starvation problems and queueing issues found in ROS1 applications. Also, ROS2 services can become synchronous.

ROS1 and ROS2 Actions

In ROS1, actions were never in the master node core functionalities. As a result, actions were integrated into ROS1 on top of ROS1 topics. This solution handled the issue that services were not asynchronous and did not have a feedback or cancelation mechanism. In ROS2, actions are now part of the ROS2 core. The Python API in ROS2 is nearly identical to the Python API in ROS1, so there are no issues with legacy code.

Underneath both ROS1 and ROS2, actions still use topics for feedback and goal status and (asynchronous) services to set a goal, cancel a goal, and request a result. ROS2 actions now have a command-line tool! You can now send an action goal to a server directly from the terminal as you do with a service.

ROS1 and ROS2 Packages

The build system in ROS1 is `catkin`. You use `catkin_make` or `catkin build` to create ROS1 packages. In ROS2, `catkin` no longer exists. Ament is the new building system, and on top of that, you get the colcon command-line tool. Use the `colcon build` command in your ROS2 workspace to compile packages quickly.

ROS1 and ROS2 Command-Line Tools

Most of the ROS1 and ROS2 command-line tools are nearly identical. In ROS1, the `rostopic list` command lists all topics, and in ROS2, `ros2 topiclist` produces the same result. In most cases, you need to remember to write `ros2` followed by the name of the tool you want to use in ROS2, as you would do in ROS1.

ROS1 and ROS2 OS Support

ROS2 is far more accessible to robotic platforms not based on Ubuntu. ROS1 only supports the Ubuntu Linux operating system (20.04 for Noetic ROS1). However, ROS2 now supports Ubuntu, macOS, and Windows 10. All of these can now operate with each other. With ROS2, the rover could be based on the Raspberry Pi 4 and Ubuntu. We could then have the ground control station use only Windows.

ROS1 (ros1_bridge) Link with ROS2

Working with a legacy ROS1 code base and needing to develop new features with ROS2 can be achieved with the ROS2 package named ros1_bridge. This creation provides communication between ROS1 and ROS2 applications, making, as its name suggests, a bridge between ROS1 and ROS2. This ROS2 package will be essential as the May 2025 end-of-life service is reached for ROS1 Noetic. Therefore, please start porting your ROS1 applications now to ROS2 systems. Please refer to the following hyperlinks for more information: https://github.com/ros2/ros1_bridge and https://index.ros.org/p/ros1_bridge/.

ROS1, Ubuntu, Raspbian, and the Raspberry Pi 4

There are three major operating systems on the Raspberry Pi 4 supporting the ROS1 Noetic system. These are the Raspbian and the Ubuntu 20.04 desktop and server operating systems. There is some need to note that the best way to determine that all of your applications work is to use an **8 GB ram version of the Raspberry Pi 4** and use only the Ubuntu 20.04 desktop or the stripped-down Ubuntu server operating system. You want to be sure that all of your applications and their libraries are fully available, as you would expect had you developed your ROS1 Noetic application on a standard Ubuntu Linux system on a PC laptop or desktop. The following is a list of references that will aid you in installing ROS1, Ubuntu, or Raspbian OS on the 8 GB Raspberry Pi 4:

```
https://roboticsbackend.com/install-ubuntu-on-raspberry-pi-
without-monitor/
https://learn.sparkfun.com/tutorials/how-to-use-remote-
desktop-on-the-raspberry-pi-with-vnc/all
https://ramith.fyi/setting-up-raspberry-pi-4-with-
ubuntu-20-04-ros-intel-realsense/
```

ROS2, Ubuntu, and Raspberry Pi 4

Only two operating systems on the Raspberry Pi 4 support the ROS2 system. These are the Ubuntu 20.04 desktop and server operating systems. Note that the best way to determine that all of your applications work correctly is to use an **8GB ram version of the Raspberry Pi 4** and use only the Ubuntu 20.04 server to maximize memory usage on the Raspberry 4 system. The following is a list of references that reveal the proper way to install ROS2 on the Raspberry Pi 4 module:

```
https://singleboardblog.com/install-ros2-on-raspberry-pi/
https://ubuntu.com/blog/ubuntu-20-04-lts-is-certified-for-
the-raspberry-pi
https://docs.ros.org/en/foxy/How-To-Guides/Installing-on-
Raspberry-Pi.html
https://medium.com/swlh/raspberry-pi-ros-2-camera-
eef8f8b94304
```

ROS1, ROS2, Raspberry Pi 4, and Rover

The fundamental idea of using the ROS environment is that once you have debugged your application on a standard Ubuntu (ROS1) or Ubuntu, Windows, or macOS (ROS2) OS, you should be able to transition your designs to an actual rover. Whichever version of ROS you decide to use (ROS1 or ROS2), and the supporting operating system, will allow you to create the brains of the rover. However, before we connect the Raspberry Pi 4 to the rover, we need to be confident that we can wirelessly connect our laptop computer to the Raspberry Pi 4 using a wireless connection via SSH connection. This step will be critical if we are to use the Rviz environment on the Ubuntu laptop or desktop to connect to the Raspberry Pi 4, acting as the central processor of the rover. The following is a list of references on the internet:

```
https://roboticsbackend.com/install-ubuntu-on-raspberry-pi-
without-monitor/
https://medium.com/@nikosmouroutis/how-to-setup-your-
raspberry-pi-and-connect-to-it-through-ssh-and-your-local-
wifi-ac53d3839be9
https://medium.com/swlh/raspberry-pi-ros-2-camera-
eef8f8b94304
https://www.hackster.io/techmirtz/connect-raspberry-pi-to-
your-laptop-screen-and-keyboard-a8a2a7
```

Summary

We have now reviewed the basic concepts to allow one to port a ROS1 system with a cognitive deep learning system to a rover platform. The Raspberry Pi 4 module will help us explore any environment (Egyptian catacombs) with genuine autonomy.

REVIEW EXERCISES

Exercise 12.1: What differences are there between ROS1 and ROS2?

Exercise 12.2: What would you need to do to make a legacy application from ROS1 work on ROS2?

Exercise 12.3: Please look up experimental algorithmics on the internet and tell us how we can use them to optimize ROS applications on the Raspberry Pi 4?

Exercise 12.4: Are command-line tools in ROS2 a good inclusion? Please explain.

## CHAPTER 13

Further Considerations

Now that you have learned about and built your own AI-enabled rover, it may be tempting to jump right in and perform the catacomb mission. However, the adage of crawling before you walk, walking before you run is applicable. The last thing you would want is to perform the mission and fail to return. Thus, testing the rover is appropriate. This chapter covers the following topics:

- Investigating and locating the cause of an accident with a land-based drone

- Understanding the future of unmanned systems and artificial intelligence

- Understanding how unmanned systems can build a better tomorrow

Designing Your First Mission

The following are the test missions under controlled conditions that you should perform with your rover:

- Manual control

- A simple corridor on flat terrain

© David Allen Blubaugh, Steven D. Harbour, Benjamin Sears, Michael J. Findler 2022
D. A. Blubaugh et al., *Intelligent Autonomous Drones with Cognitive Deep Learning*,
https://doi.org/10.1007/978-1-4842-6803-2_13

- Complex-shaped corridor with uneven terrain

- Complex corridor with uneven terrain and obstacles

- Additional testing as required

Each of these missions will allow for isolating and testing specific functions of the rover. Conducting these tests will reduce the risk of a failed mission, as errors in programming, design, build, and other unknown factors can be corrected. Each of these testing missions should be performed in an open area. This allows for the rover to be recovered or aborted in the event of an issue.

Manual Control

This mission is straightforward and is only applicable if you have installed a remote-control receiver on the vehicle. Simply put, drive the rover around. Test out all of the drive controls of the vehicle. We call this a controllability assessment. By going forward, reversing, and turning left and right you will be able to identify if there are any issues in the wiring of the rover.

Simple Corridor on Flat Terrain

This mission will test the sensors and the AI. Create a straight corridor out of cardboard, or use a hallway. Using the interface program, command the rover to start at one end of the hallway and maneuver to the end of the hallway. This mission tests out the ability of the drive system and the AI to work together.

Complex-shaped Corridor with Uneven Terrain

This mission will further test the sensors and the AI. Using the corridor you created for the previous mission, add cardboard to the sides to create an uneven boundary wall and uneven terrain. The cardboard added to the walls needs to vary, allowing for twists and turns. This can be in the form of carpeting, ramps, and so on. Once again, program the rover to start at one end and maneuver to the other.

Complex Open Corridor with Uneven Terrain and Obstacles

This last testing mission integrates the rover's many issues in the catacombs. Modify the corridor you created for the previous mission with further twists and turns and add more complex terrain. This mission also incorporates obstacles that the rover will have to maneuver around. Place objects in the way of the rover. Once again, program the rover to start at one end and maneuver to the other. The success of this final mission means you have a fully functional rover!

Additional Testing as Required

Additional testing may be warranted if you have opted to install different sensors. Test each additional sensor individually. This will allow you to isolate any issues that might arise. If there is a problem, conducting all of these tests at once will make it difficult to separate the specific sub-system that is causing the issue.

What to Do if the AI Rover Crashes

There are many issues the rover might encounter that could cause the system to crash. A crash can consist of a software crash or a physical crash into an obstacle. We rely on a system of elements to work together without a human in the loop to take control to prevent a collision. The key is to collect data from a crash to build a better AI-enabled rover. Some problems can be corrected in the software. Others are hardware related. For instance, if one of the drive motors fails, then the rover will drive in a circle. If the system crashes, then we need to analyze the data to pinpoint the problem.

The Pixhawk autopilot automatically creates a log file and is stored in non-volatile memory. This can be downloaded from the autopilot and loaded into the open-source Mission Planner software or other software designed to analyze the logs to isolate the problem. Both automated log scanning and manual log scanning will allow you to troubleshoot the issues. This troubleshooting covers the primary drive system.

Mission Ideas

The mission this book centers on is exploring an unexplored catacomb in Egypt. This is not the only mission you could program your AI rover to perform. The following sections provide a few examples of missions you could perform with your rover.

Zombie Hunter

In this mission, you will program your AI-enabled rover to identify the undead in a desolate wasteland and map their locations using your newfound Geographic Information System (GIS) skills. To perform this mission, the rover's AI must use the sense-and-avoid system to identify zombies from humans. Specifically, you must train the AI to classify humans from zombies, and then log their location in the GIS software. To confirm your programing is successful, print out small images of zombies and humans no taller than the rover itself and set these up throughout your corridor. Start your mission at one end. Your task will be successful if your rover makes it to the end of the corridor and correctly identifies and maps the zombies and humans.

Home Delivery

This mission is as simple as it sounds. The rover is programmed to use its SLAM sensors to deliver a payload to other areas of your house. One consideration with this mission is the payload capacity of the rover. Exceeding this capacity will lead to a reduced range or overall mission failure.

Home Security

For this mission, you program the rover to patrol your house. For any detected motion, the rover could trigger an alarm and record images and/or video. You will need to consider thresholds of activity to prevent false positives.

Other Missions

Your AI rover can be used for many different missions. Your imagination is the only limiting factor. Continue to use the testing and evaluation methods to confirm your system is programmed and fully operational before trying a live mission.

Like It or Not, We Now Live in the Age of Skynet

Technology is rapidly advancing. What was once thought of as science fiction is quickly becoming a reality. In 1865, Jules Verne's science fiction novel *From the Earth to the Moon* spoke of launching humans to the moon via a cannon. Little did he know that 104 years later, humans would actually land on the moon. Today, we have rovers and unmanned aircraft systems (UAS) on Mars, *Voyager* venturing outside of the solar system, and telescopes on Earth and in space that can gather massive amounts of data. Science fiction is quickly becoming a reality in many industries. We have self-driving cars, access to robots for all sorts of applications, and access to collective knowledge via the internet. Could Skynet be created and the doomsday scenario it presents in the Terminator franchise come to be? The answer is yes, it is quite possible. It is up to AI designers and programmers to prevent this from happening.

Future Battlefields and Skies Will Have Unmanned Systems

Since the creation and adoption of unmanned systems on the battlefield, many in the defense industry have been researching and developing systems that can push the envelope further. UAS used to be used solely for intelligence, surveillance, and reconnaissance. They are piercing the fog of war with myriad sensors. This gives battlefield commanders and soldiers real-time access to the battlefield. They are allowing for decisions to be made to win the conflict. Other robotic systems have been in development. One rover system called the Multi-Function Utility/Logistics and Equipment Vehicle (MULE) was designed with three variants. The variants of this MULE were transport, countermine, and light assault. This program was ultimately canceled by the military, yet highlights the fact that robotic systems can serve on the battlefield in many different ways. Uncrewed robotic systems serve a need from air, sea, and ground. This need is to remove the human element from danger. If an uncrewed robotic tank gets destroyed in battle, the loss is of the monetary value of the tank, not the human piloting the tank. All the training and experience gathered by the talk operator is not lost. All they need to do is connect to another tank and keep fighting. Battles of the future would not be as costly in terms of life, but instead in monetary value. The winner of these battles will be those that have a deep manufacturing base that can keep producing these robotic systems, as well as logistics and repair locations to support those that are in battle.

Necessary Countermeasures

The development of a new technology produces a counter-industry. A perfect case in point is personal computers. These systems became affordable in the 1980s and 1990s, and coupled with ease of access to the internet, hackers and those with ill intent created viruses and malware. To protect users, a new industry was born, one that constantly battles viruses and protects computer systems from hackers. As soon as the antivirus software has identified, blocked, and removed a virus, a new one is written, or a new exploit is found. This has been a constant battle and one that will most likely never end!

Countermeasures for all unmanned systems are being developed and deployed. The current hot zone in this counter-industry is with aircraft systems. It was quickly realized that anyone could fly these systems into sensitive or restricted areas; for instance, the White House, military bases, power plants, etc. Thus, a counter-industry was born.

The counter-UAS industry is looking at many different ways of stopping or identifying UAS. Some of the many systems being developed or deployed are listed here:

- Radar identification and localization

- Acoustic identification and localization

- Infrared identification

- Combination of identification sensors

- Net capture

- Frequency jamming

- GNSS jamming or spoofing

- Microwave pulse

- UAS vs. UAS to intercept and capture

- Broadcast-system kill codes

- And many more in development

This book seeks to meld unmanned systems and AI together. There is already a budding counter-industry for unmanned systems. However, currently, AI is still relatively new. Many in the defense industry are thinking about this counter-AI issue and developing a cohesive strategy. Whatever this strategy looks like, it is nevertheless secret.

Final Considerations for More Advanced AI-Enabled

Drones

Technology has dramatically improved humans' lives. AI and unmanned systems are just another technology that will enhance humanity's quality of life. Someday, humans will never learn to drive cars, and will have robotic household robots that will do all the cleaning, laundry, and dishes. With all the good these two technologies can produce, there will be bad intentions. Counter-strategies are being developed. Nevertheless, AI and unmanned systems are here to stay.

Summary

The significant points of this chapter are to always test your system in a controlled environment. This prevents the total loss of the rover. System crashes and physical crashes are inevitable. Practice analyzing log files in Mission Planner and the AI system to isolate the problem. Reprogram your rover for different missions.

References

"Welcome To The QGIS Project!" (2022). Qgis.Org. `https://www.qgis.org/en/site/`.

"Github - Locusrobotics/Qgis_Ros: ROS QGIS Plugin Prototype." (2022). Github. `https://github.com/locusrobotics/qgis_ros`.

"Roscon 2018 Madrid: Unleashing The GIS Toolbox On Real-Time Robotics." (2022). Vimeo. `https://vimeo.com/293539252`.

"QGIS Training Manual — QGIS Documentation Documentation." (2022). Docs.Qgis.Org. https://docs.qgis.org/3.10/en/docs/training_manual/index.html.

"Introduction To GIS Fundamentals - NYC DOT." (2022). Training. Datapolitan.Com. https://training.datapolitan.com/qgis-training/Introduction_to_GIS_Fundamentals/#1.

"Rover Home — Rover Documentation." (2022). Ardupilot.Org. https://ardupilot.org/rover/index.html.

"When Problems Arise — Rover Documentation." (2022). Ardupilot. Org. https://ardupilot.org/rover/docs/common-when-problems-arise.html.

"Logs — Rover Documentation." (2022). Ardupilot.Org. https://ardupilot.org/rover/docs/common-logs.html.

"Multifunctional Utility/Logistics and Equipment (MULE) Vehicle Will Improve Soldier Mobility, Survivability and Lethality." (2022). Asc. Army.Mil. https://asc.army.mil/docs/pubs/alt/2008/2_AprMayJun/articles/27_Multifunctional_Utility-Logistics_and_Equipment_(MULE)_Vehicle_Will_Improve_Soldier_Mobility,_Survivability_and_Lethality_200804.pdf.

APPENDIX A

Bayesian Deep Learning

Handling uncertainty is what Bayesian networks are good at. Bayesian Gaussian mixture models use the `BayesianGaussianMixture` class, which is capable of giving weights equal (or close) to zero to remove unnecessary clusters, rather than your manually searching for the optimal number of clusters. This appendix provides a crash course on the essentials of Bayesian networks.

Bayesian Networks at a Glance

Set the number of clusters `n_components` to a value that you have good reason to believe is greater than the optimal number of clusters (this assumes some minimal knowledge about the problem at hand), and the algorithm will eliminate the unnecessary clusters automatically.

For example, let's set the number of clusters to 10 and see what happens:

```
>>> from sklearn.mixture import BayesianGaussianMixture
>>> bgm = BayesianGaussianMixture(n_components=10, n_init=10,
random_state=42) >>> bgm.fit(X) >>> np.round(bgm.weights_, 2)
array([0.4 , 0.21, 0.4 , 0. , 0. , 0. , 0. , 0. , 0. , 0. ])
```

Perfect: The algorithm automatically detected that only three clusters are needed.

© David Allen Blubaugh, Steven D. Harbour, Benjamin Sears, Michael J. Findler 2022
D. A. Blubaugh et al., *Intelligent Autonomous Drones with Cognitive Deep Learning*,
https://doi.org/10.1007/978-1-4842-6803-2

In this model, the cluster parameters (including the weights, means, and covariance matrices) are not treated as fixed model parameters, but rather as latent random variables, like the cluster assignments (see Figure A-1). So z now includes both the cluster parameters and the cluster assignments.

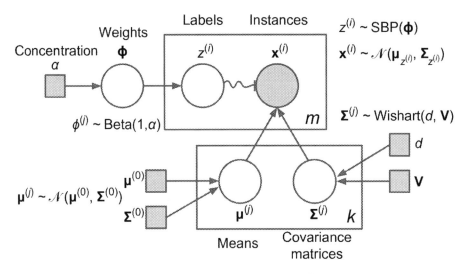

Figure A-1. *Bayesian Gaussian mixture model*

What Are You? Two Camps . . .

The *frequentist* definition sees probability as the long-run expected frequency of occurrence. $P(A) = n/N$, where n is the number of times event A occurs in N opportunities.

"probability" = long-run fraction having this characteristic.

The *Bayesian* view of probability is related to degree of belief. It is a measure of the plausibility of an event given incomplete knowledge. They say that only the data are real. The population mean is an abstraction, and as such some values are more believable than others based on the data and their prior beliefs.

"probability" = degree of believability.

Bayesian Decision Theory

- Fundamental statistical approach to classification
- Quantifies tradeoffs between classification decisions
  - Probabilities
  - Costs of decisions
- Requires:
  - Known *a priori* probabilities
  - Prior probabilities
  - Likelihood PDFs

$$P(c_i|x) = \frac{p(x|c_i)P(c_i)}{p(x)}$$

Bayes Theorem

Bayesian belief networks (BBNs) are probabilistic graphical models (PGMs) that represent conditional dependencies between random variables.

BBN Conditional Property

Conditional probability is the probability of a random variable's occurring when some other random variable is given. It is shown by

$$P(X|Y)$$

If these two random variables are dependent, then

$$P(X|Y) = \frac{P(X,Y)}{P(Y)}$$

If they are independent, then

$$P(X|Y) = P(X)$$

Mathematical Definition of Belief Networks

The probabilities are calculated in belief networks by the following formula:

$$P(X_1,\ldots,X_N) = \prod_{i=1}^{N} P(X_i|Parents(X_i))$$

As you can see from the formula, to calculate the joint distribution we need to have conditional probabilities indicated by the network (Listing A-1). But further, if we have the joint distribution, then we can start to ask interesting questions.

Listing A-1. Python Example of Belief Network

```
# create the nodes
season = BbnNode(Variable(0, 'season', ['winter', 'summer']),
[0.5, 0.5])
atmos_pres = BbnNode(Variable(1, 'atmos_press', ['high',
'low']), [0.5, 0.5])
allergies = BbnNode(Variable(2, 'allergies', ['allergic', 'non_
alergic']), [0.7, 0.3, 0.2, 0.8])
```

```
rain = BbnNode(Variable(3, 'rain', ['rainy', 'sunny']), [0.9,
0.1, 0.7, 0.3, 0.3, 0.7, 0.1, 0.9])
grass = BbnNode(Variable(4, 'grass', ['grass', 'no_grass']),
[0.8, 0.2, 0.3, 0.7])
umbrellas = BbnNode(Variable(5, 'umbrellas', ['on', 'off']),
[0.99, 0.01, 0.80, 0.20, 0.20, 0.80, 0.01, 0.99])
dog_bark = BbnNode(Variable(6, 'dog_bark', ['bark', 'not_
bark']), [0.8, 0.2, 0.1, 0.9])
cat_mood = BbnNode(Variable(7, 'cat_mood', ['good', 'bad']),
[0.05, 0.95, 0.95, 0.05])
cat_hide = BbnNode(Variable(8, 'cat_hide', ['hide', 'show']),
[0.20, 0.80, 0.95, 0.05, 0.95, 0.05, 0.70, 0.30])bbn = Bbn() \
    .add_node(season) \
    .add_node(atmos_pres) \
    .add_node(allergies) \
    .add_node(rain) \
    .add_node(grass) \
    .add_node(umbrellas) \
    .add_node(dog_bark) \
    .add_node(cat_mood) \
    .add_node(cat_hide) \
    .add_edge(Edge(season, allergies, EdgeType.DIRECTED)) \
    .add_edge(Edge(season, umbrellas, EdgeType.DIRECTED)) \
    .add_edge(Edge(season, rain, EdgeType.DIRECTED)) \
    .add_edge(Edge(atmos_pres, rain, EdgeType.DIRECTED)) \
    .add_edge(Edge(rain, grass, EdgeType.DIRECTED)) \
    .add_edge(Edge(rain, umbrellas, EdgeType.DIRECTED)) \
    .add_edge(Edge(rain, dog_bark, EdgeType.DIRECTED)) \
    .add_edge(Edge(rain, cat_mood, EdgeType.DIRECTED)) \
    .add_edge(Edge(dog_bark, cat_hide, EdgeType.DIRECTED)) \
    .add_edge(Edge(cat_mood, cat_hide, EdgeType.DIRECTED))
```

```
with warnings.catch_warnings():
    warnings.simplefilter('ignore')

    graph = convert_for_drawing(bbn)
    pos = nx.nx_agraph.graphviz_layout(graph, prog='neato')
    plt.figure(figsize=(20, 10))
    plt.subplot(121)
    labels = dict([(k, node.variable.name) for k, node in bbn.
    nodes.items()])
    nx.draw(graph, pos=pos, with_labels=True, labels=labels)
    plt.title('BBN DAG')
```

Summary

References

Chang, T.H., Hsu, C.S., Wang, C., & Yang, L.-K. (2008). On board measurement and warning module for measurement and irregular behavior. *IEEE Transactions on Intelligent Transportation Systems, 9*(3), 501–513.

Clark, J.D., & Harbour, S.D. (2019). Unpublished.

Clark, J.D., Mitchell, W.D., Vemuru, K.V., & Harbour, S.D. (2019). Unpublished.

Dayan, P., Abbott, L.F., & Abbott, L. (2001). Theoretical neuroscience: computational and mathematical modeling of neural systems.

Friston, K., & Buzsáki, G. (2016). The functional anatomy of time: what and when in the brain. *Trends in cognitive sciences, 20*(7), 500–511.

Friston K. (2018). Am I Self-Conscious? (Or Does Self-Organization Entail Self-Consciousness?). *Frontiers in psychology, 9*, 579. doi:10.3389/fpsyg.2018.00579

Gerstner, W., & Kistler, W. (2002). *Spiking Neuron Models: Single Neurons, Populations, Plasticity*. Cambridge University Press.

Harbour, S.D., & Christensen, J.C. (2015, May). A neuroergonomic quasi-experiment: Predictors of situation awareness. In Display Technologies and Applications for Defense, Security, and Avionics IX; and Head-and Helmet-Mounted Displays XX (Vol. 9470, p. 94700G). SPIE.

Harbour, S.D., Rogers, S.K., Christensen, J.C., & Szathmary, K.J. (2015, 2019). Theory: Solutions toward autonomy and the connection to situation awareness. Presentation at the 4th Annual Ohio UAS Conference. Convention Center, Dayton, Ohio. USAF.

Harbour, S.D., Clark, J.D., Mitchell, W.D., & Vemuru, K.V. (2019). Machine Awareness. *20th International Symposium on Aviation Psychology*, 480–485. https://corescholar.libraries.wright.edu/isap_2019/81

Kidd, C., & Hayden, B.Y. (2015). The Psychology and Neuroscience of Curiosity. *Neuron, 88*(3), 449–460.

Kistan, T., Gardi, A., & Sabatini, R. (2018). Machine learning and cognitive ergonomics in air traffic management: Recent developments and considerations for certification. *Aerospace, 5*(4), article no. 103.

Loewenstein G. (1994). The Psychology of Curiosity: A Review and Reintrepretation. *Psychological Bulletin, 116*(1), 75–98.

Mitchell, W.D. (February, 2019), private communication.

Murphy, R.R. (2019). *Introduction to AI robotics*. MIT press.

Rogers, S. (2019). Unpublished.

Sharpee, T.O., Calhoun, A.J., & Chalasani, S.H. (2014). Information theory of adaptation in neurons, behavior, and mood. *Current opinion in neurobiology*, 25, 47–53.

Li, L.S., Hansman, R.J., Palacios, R., and Welsch, R. (2016). Anomaly detection via Gaussian mixture model for flight operation and safety monitoring, *Transportation Technologies, Part C: Emerging Technologies, 64*, 45–57.

Pour, A.G., Taheri, A., Alemi, M., & Meghdari, A. (2018). Human-Robot facial expression reciprocal interaction platform: Case studies on children with autism. *International Journal of Social Robotics, 10*(2), 179–198.

Vemuru, K.V., Harbour, S.D., & Clark, J.D. (2019). Reinforcement Learning in Aviation, Either Unmanned or Manned, with an Injection of AI. *20th International Symposium on Aviation Psychology*, 492–497. https://corescholar.libraries.wright.edu/isap_2019/83

Xu, S.T., Tan, W.Q., Efremov, A.V., Sun, L.G., & Qu, X. (2017). Review of control models for human pilot behavior, *Annual Review in Control, 44*, 274–291.

Zhao, W.Z., He, F., Li, L.S., & Xiao, G. (2018). An adaptive online learning model for flight data cluster analysis, In *Proc. of 2018 IEEE/AIAA 37th Digital Avionics Systems Conference, IEEE-AIAA Avionics Systems Conference* (pp.1–7). London, UK.

APPENDIX B

OpenAI Gym

OpenAI Gym is open to anyone interested in machine learning (ML) and artificial intelligence (AI) research. It was founded in 2015 by Elon Musk, Sam Altman, and others, who collectively pledged $1 billion. The idea was to make this type of research more available in the area of AI, with the stated goal of promoting and developing amicable AI in a way that benefits humanity as a whole (`https://en.wikipedia.org/wiki/OpenAI`).

Reinforcement learning (RL) is the machine learning methodology that involves creating sequences of decisions. RL encompasses a mathematical theory and has found a range of sensible applications. Recent advances that mix deep learning with reinforcement learning have led to a good deal of optimism in the field. It has become evident that general algorithms like policy gradients and Q-learning will achieve innovative performance on challenging issues, while not involving problem-specific engineering. To build on recent progress in reinforcement learning, the analysis community desires innovative benchmarks via which to check algorithms. A range of benchmarks have been discharged, like the Arcade Learning setting (ALE), which exposed a set of Atari 2600 games as reinforcement learning problems, and recently the RL Lab benchmark for continuous management, to which we tend to refer the reader for a survey of alternative RL benchmarks. Open AI Gym aims to mix the most effective components of those previous benchmark collections in a software package that's convenient and accessible. It includes various

© David Allen Blubaugh, Steven D. Harbour, Benjamin Sears, Michael J. Findler 2022
D. A. Blubaugh et al., *Intelligent Autonomous Drones with Cognitive Deep Learning*,
https://doi.org/10.1007/978-1-4842-6803-2

tasks (called environments) with a familiar interface, and can grow over time. The environments are versioned in a manner that will ensure that results stay meaningful and reproducible, because the software package is regularly updated. Alongside the software package library, OpenAI Gym encompasses a website (`gym.openai.com`), where one can view scoreboards for all environments that showcase results submitted by users. Users are encouraged to supply links to ASCII text files and directions on how to reproduce their results. Reinforcement learning assumes that an agent is placed in a setting. At each step, the agent takes an action, and it receives an observation and reward from the setting (see "Train Your Reinforcement Learning Agents at the OpenAI Gym" at `https://developer.nvidia.com/blog/train-reinforcement-learning-agents-openai-gym/`). An RL formula maximizes some life of the agent's total reward because the agent interacts with the setting. In the RL literature, the setting is formalized as a part-evident Markov call method (POMDP).

OpenAI Gym focuses on the episodic setting of reinforcement learning, where the agent's expertise is dampened into a series of episodes. In every episode, the agent's initial state is indiscriminately sampled from a distribution, and therefore the interaction yields until the setting reaches a terminal state. The goal in episodic reinforcement learning is to maximize the expectation of total reward per episode and to attain a high level of performance in as few episodes as is doable. The following code snippet shows one episode with a hundred timesteps. It assumes that there's an object called agent that takes an observation at every timestep, and an object known as env that is that the. The design of OpenAI Gym relies on the authors' expertise in developing and examining reinforcement learning algorithms, and our expertise with and mistreatment by previous benchmark collections. In this appendix, we are going to summarize a number of our style decisions. Environments, not agents. Two core ideas are the agent and, therefore, the setting.

We have chosen to solely offer an abstraction for the setting, and not for the agent. This alternative maximizes convenience for users and permits them to implement completely different types of agent interface. First, one might imagine an "online learning" vogue, where the agent takes (`observation, reward, done`) as an input at every timestep and performs learning updates incrementally. An agent is named, with the observation used as input in an alternate "batch update" vogue. Therefore, the reward information is collected individually by the RL formula, and later it is the accustomed reason for an update. By solely specifying the agent interface, we tend to permit users to put in writing their agents with either of those designs. Emphasize sample quality, not simply final performance. The performance of an RL formula on an environment may be measured on two axes: first, the ultimate performance; second, the number of times it takes to learn—the sample quality. To be more specific, final performance refers to the typical reward per episode after learning is complete. Learning time may be measured in multiple ways. One is to count the number of episodes before an intensity level of average performance is exceeded. RL presents new challenges for benchmarking. Performance is measured within the supervised learning setting by looking at prediction accuracy on a check set, where the correct outputs are hidden from contestants. In RL, it is less straightforward to assess performance, except by running the users' code on a set of unseen environments, which might be computationally expensive. While not a hidden check set, one should certify that an algorithm failed to "overfit" on its tested issues (for example, through parameter tuning). We prefer to encourage a referee method for deciphering results submitted by users.

Thus, OpenAI Gym asks users to form. OpenAI Gym contains a set of environments (POMDPs) that can grow over time. The following environments were included at the time of the gym's initial beta launch:

- Classic management text: small-scale tasks from RL literature

- Algorithmic: perform computations like adding multi-digit numbers and reversing sequences. Most of these tasks need memory, and varied sequence length may choose their issue.

- Atari: classic Atari games, with screen pictures or RAM as input; mistreatment of the Arcade Learning setting

- Board games: Presently, we have included the sport of enduring 9x9 and 19x19 boards, wherever the Pachi engine In the future, we hope to increase OpenAI Gym in many ways that

- Multi-agent setting: It will be attention-grabbing to eventually embody tasks on which agents should collaborate or compete with other agents.

- Course of study and transfer learning: Right now, the tasks are meant to be solved from scratch. Later, it will be more attention-grabbing to contemplate sequences of tasks, so the formula is trained on one task when the opposite (OpenAI Gym – arXiv Vanity: `https://www.arxiv-vanity.com/papers/1606.01540/`). Here, we will produce sequences of more and more tough tasks that are meant to be solved so as.

- Real-world operation: Eventually, we might prefer to integrate the gym API with robotic hardware, validating reinforcement learning algorithms out in the world.

Getting Started with OpenAI Gym

The gym is a toolkit for developing and comparing reinforcement learning algorithms. It makes no assumptions concerning the structure of your agent and is compatible with any numerical computation library, like TensorFlow or Theano.

The OpenAI Gym library is an assortment of environments that you can use to figure out reinforcement learning algorithms. These environments have a shared interface, permitting you to jot down general algorithms.

Installation

To get started, you must install Python 3.5+. Install OpenAI Gym utilizing pip as follows:

```
pip
install pip install gym
```

Moreover, you are wise to go!

Building from supply If

If you want, you can clone the gym repository directly. This can be helpful once you are modifying the gym code itself or adding environments. Transfer and install using ("Getting Started with Gym – OpenAI" at https://gym.openai.com/docs/):

```
git
clone https://github.com/openai/gym
cd gym pip install -e.
```

You can later run `pip install -e.[all]` to perform a complete installation containing all environments. This needs a lot of related dependencies, together with CMake and a recent pip version. (See "Getting Started with Gym – OpenAI" at `https://gym.openai.com/docs/`)

Environments

Here's a minimum example of getting one thing running. This can run an associate instance of the CartPole-v0 atmosphere for a thousand timesteps, rendering the atmosphere at every step. You ought to see a window pop up generating the classic cart-pole problem (See "Getting Started with Gym – OpenAI" at `https://gym.openai.com/docs/`):

```
Import gym env = gym.make('CartPole-v0')
env.reset() for
_ in range(1000):

  env.render()

  env.step(env.action_space.sample()) # take a random action
  env.close()
```

It Ought to Look Like This

Typically, we will finish the simulation before the cart pole can travel off screen. Please ignore the warning concerning `step()`, even supposing this atmosphere has already come back with `done = True`.

If you wish to see another environment in action, strive for commutation CartPole-v0 higher than with one thing, like MountainCar-v0, MsPacman-v0 (requires the Atari dependency), or Hopper-v1 (requires the MuJoCo dependencies). Environments all descend from the `Env` base category.

> **Note**   If you are missing any dependencies, see "Getting Started with Gym – OpenAI" at `https://gym.openai.com/docs/`.

You will get a helpful error message telling you what you are missing. (A dependency might give you a hard time by not providing clear instructions on how to fix it.) Putting in a missing dependency is usually pretty straightforward. You will need a MuJoCo license for Hopper-v1.

Observations

If you ever wish to do more than take random actions at every step, it'd be wise to understand what your efforts do to the atmosphere.

The environment's step returns precisely what we want. The action returns four values. The first is observation (object): an associated atmosphere-specific object representing your observation of the environment. For instance, component information from a camera, joint angles, and joint velocities of a (see "Getting Started with Gym – OpenAI" at `https://gym.openai.com/docs/golem`). Or even the board state in an exceedingly parlor game (See Getting Started with Gym – OpenAI" at `https://gym.openai.com/docs/`). The second is Reward (float): quantity of reward achieved by the previous action. The size varies between environments; however, the goal is often to extend your total reward. The third is Done (Boolean): whether or not it's time to reset the atmosphere. Most (but not all) tasks' area units are divided into well-defined episodes, and done being True indicates the episode has terminated. (For example, maybe the pole tipped too far so you lost your last life.) The fourth is Info (dict): diagnostic information helpful for debugging. It will typically benefit learning (for example, it would contain the raw possibilities behind the environment's last state change). However, official evaluations of your agent do not seem to be allowed to use this for learning.

This is an associate implementation of the classic "agent–environment loop." Every timestep, the agent chooses an associated action, and the atmosphere returns an associated observation and a gift.

Job reset () starts the process, which returns the initial associated observation.

Thus a lot of correct methods to write the previous code would have to respect the done flag:

```
import gym
env
= gym.make('CartPole-v0')
for
i_episode in range(20):

    observation = env.reset()

    for t in range(100):

        env.render()

        print(observation)

        action = env.action_space.sample()

        observation, reward, done, info = env.step(action)

        if done:

            print("Episode finished once timesteps".format(t+1))
            break
env.close()
```

This should provide a video and output just like the following. You ought to be able to see wherever the resets happen.

```
[-0.061586 -0.75893141 zero.05793238 one.15547541]
[-0.07676463 -0.95475889 zero.08104189 one.46574644]
[-0.0958598 -1.15077434 zero.11035682 one.78260485]
[-0.11887529 -0.95705275 zero.14600892 one.5261692 ]
[-0.13801635 -0.7639636 zero.1765323 one.28239155]
[-0.15329562 -0.57147373 zero.20218013 one.04977545]
The episode finished after fourteen timesteps
[-0.02786724 zero.00361763 -0.03938967 -0.01611184]
[-0.02779488 -0.19091794 -0.03971191 zero.26388759]
[-0.03161324 zero.00474768 -0.03443415 -0.04105167]
```

Spaces

We've been sampling random actions from the environment's auction house in the higher-than-ten examples. However, what is the area unit of those actions? Each atmosphere comes with an `action_space` that is associated with an `observation_space`. These attributes have an area unit of type house, and they describe the format of valid actions and observations, as follows:

```
import
gym
env
= gym.make('CartPole-v0')
print(env.action_space)
#>
Discrete(2)
print(env.observation_space)
#>
Box(4,)
```

The distinct house permits a hard and fast variety of non-negative numbers; thus, valid action units are either zero or one in this case. The Box house represents an n-dimensional associated box; thus, valid observations are an array of four numbers. We will check the box's bounds:

```
print(env.observation_space.high)
#>
array([ two.4 , inf, 0.20943951, inf])
print(env.observation_space.low)
#>
array([-2.4 , -inf, -0.20943951, -inf])
```

It will be useful to jot down generic code that works for several different environments. Box and distinct area units are the main common areas. You might need to sample from an area or ensure something belongs to it.

From gym, import areas as follows:

```
space
= houses.Discrete(8) # Set with eight components
x
= space.sample()
assert
house.contains(x)
assert
house.n == 8
```

For Carole-v0, one of the actions applies force to the left, and one of them applies force to the right. (Can you work out which is which?)

Fortunately, the higher your learning formula, the less you will need to attempt to interpret these numbers yourself.

Available Environments

OpenAI Gym comes with myriad environments that vary from straightforward to troublesome and involve many different sorts of information. Review the complete list of domains to get a bird's-eye view.

Classic management and toy text: Complete small-scale tasks, mainly from the RL literature. They are here to get you started.

Algorithmic: Perform computations like adding multi-digit numbers and reversing sequences. These tasks have the extraordinary property that it is easy to vary the issue by varying the sequence length. One may object that these tasks are unit straightforward for a PC. The challenge is to find these algorithms strictly from examples.

Atari: Play classic Atari games. We have integrated the Arcade Learning atmosphere (which has had a giant impact on reinforcement learning research) into associated easy-to-install kind.

Enclosed area unit some environments from a recent benchmark by University of California–Berkeley researchers (who incidentally are connexon America this summer).

2D and 3D golems: Management of a robot in simulation. These tasks use the MuJoCo physics engine (See "Strengthening the GYM of Learning – Programmer Sought" at https://www.programmersought.com/article/90548893021/), which was designed for quick and accurate golem simulations. MuJoCo is a proprietary software package; however, it offers free trial licenses.

The written record gym's primary purpose is to supply an outsized assortment of environments that expose a standard interface and area unit versioned to permit comparisons. To list the environments on the market in your installation, raise athletic facility.envs.registry:

```
From athletic facility import envs
print(envs.registry.all())
#> [EnvSpec(DoubleDunk-v0), EnvSpec(InvertedDoublePendulum-v0),
```

EnvSpec(BeamRider-v0), EnvSpec(Phoenix-ram-v0), EnvSpec(Asterix-v0),
EnvSpec(TimePilot-v0), EnvSpec(Alien-v0), EnvSpec(Robotank-ram-v0), EnvSpec(CartPole-v0),
EnvSpec(Berzerk-v0), EnvSpec(Berzerk-ram-v0), EnvSpec(Gopher-ram-v0), ...

This will provide you with an inventory of EnvSpec objects. These outline parameters for a specific task, together with variety|the amount|the quantity} of trials to run and also the number of steps. For instance, EnvSpec(Hopper-v1) defines an associate atmosphere wherever the goal is to induce a second simulated golem to hop; EnvSpec(Go9x9-v0) defines a board game on a 9x9 board.

These atmosphere ID area units are treated as opaque strings. To make sure there are valid comparisons for the longer term, environments can never be modified in a fashion that affects performance; they are simply replaced by newer versions. We tend to suffix every atmosphere with a v0 so that future replacements will naturally be known as v1, v2, etc.

It's straightforward to feature your environments in the written record and put them on the market to run gym.make(): register() at load time.

Background

Why OpenAI Gym? (2016) Reinforcement learning (RL) is the subfield of machine learning that involves higher cognitive processes and control. It studies. However, associated agents will learn how to attain goals in an exceedingly advanced, uncertain atmosphere. It's exciting for two reasons.

RL is incredibly general, encompassing all issues that involve creating a sequence of decisions; for instance, control a robot's motors so that it is able to run and jump; making business selections like ratings and inventory management; or taking part in video games and board games. RL will even be applied to supervised learning issues with successive or structured outputs.

RL algorithms have begun to achieve good things in several troublesome environments. RL has a long history; it needed lots of problem-specific engineering until recent advances in deep learning. DeepMind's Atari results, BRETT from Pieter Abbeel's cluster, and AlphaGo all used deep RL algorithms.

First, the need for higher benchmarks. In supervised learning, progress has been driven by giant labeled datasets like ImageNet. The nearest equivalent would be an outsized and varied assortment of environments in RL. However, the present ASCII text file collections of RL environments do not have enough variety, and the area unit is typically difficult to even come upon and use.

Second, there is the lack of standardization of the environments utilized in publications. Minute variations within the underlying definition, like the reward, performance, or set of actions, will drastically alter a task's problem. This issue makes it difficult to perform printed analyses and compare results from totally different papers.

The OpenAI Gym is an endeavor to mend each issue.

Drone Gym Environment

This repository contains a pip package that is an OpenAI Gym environment for a drone that learns via RL. It also introduces the concept of interactive reinforcement learning.

Install OpenAI Gym

Install this package via `pip install -e .`

Then, make the environment as follows:

```
import gym
import gym_pull
gym_pull.pull('github.com/jnc96/drone-gym')
env = gym.make('Drone-v0')
```

See `https://github.com/matthiasplappert/keras-rl/tree/master/examples` for some examples.

Dependencies

The entire ecosystem heavily depends on TensorForce (`https://github.com/tensorforce`). OpenAI Gym was also used in the creation of the environment (`https://gym.openai.com/`).

Special thanks to Alexander Kuhnle for his help in developing this.

The Environment

The environment leverages the framework as defined by OpenAI Gym to create a custom environment. The environment contains a grid of terrain gradient values. The reward of the environment is predicted coverage, which is calculated as a linear function of the actions taken by the agent.

In Real Life

The main purpose of this entire system is to investigate how human interaction can affect the traditional reinforcement learning framework. Custom scripts were written to facilitate this, and several TensorForce scripts were also modified. These can be found in the custom scripts folder, which needs to be manually extracted and placed in the TensorForce package directory.

It was created by Jia Ning Choo in 2019 (`https://github.com/jnc96`).

Note   This repository's main branch is actively developed, please git pull frequently and feel free to open new issues (`https://github.com/utiasDSL/gym-pybullet-drones/issues`) for any undesired, unexpected, or (presumably) incorrect behavior. Thanks!

---

gym-pybullet-drones

This is a simple (`https://en.wikipedia.org/wiki/KISS_principle`) OpenAI Gym environment based on PyBullet (`https://github.com/bulletphysics/bullet3`) and used for multi-agent reinforcement learning with quadrotors.

The default DroneModel.CF2X dynamics are based on Bitcraze's Crazyflie 2.x nano-quadrotor (`https://www.bitcraze.io/documentation/hardware/crazyflie_2_1/crazyflie_2_1-datasheet.pdf`).

Everything after a $ is entered on a terminal, everything after >>> is passed to a Python interpreter.

To better understand how the PyBullet backend works, refer to its Quickstart Guide.

Suggestions and corrections are very welcome in the form of issues (`https://github.com/utiasDSL/gym-pybullet-drones/issues`) and pull requests (`https://github.com/utiasDSL/gym-pybullet-drones/pulls`), respectively.

Why Reinforcement Learning of Quadrotor Control?

A lot of recent RL research for continuous actions has focused on policy gradient algorithms and actor–critic architectures (`https://lilianweng.github.io/lil-log/2018/04/08/policy-gradient-algorithms.html`). A quadrotor is (i) an easy-to-understand mobile robot platform whose (ii) control can be framed as a continuous states and actions problem, but that, beyond one dimension, (iii) adds the complexity that in many candidate policies leads to unrecoverable states, violating the assumption of the existence of a stationary-state distribution on the entailed Markov chain.

Overview

gym-pybullet-drones
   AirSim (`https://github.com/microsoft/AirSim`)
   Flightmare (`https://github.com/uzh-rpg/flightmare`)

Physics
PyBullet
FastPhysicsEngine/PhysX
Ad hoc/Gazebo

Rendering
PyBullet
Unreal Engine 4
Unity

Language
Python
C++/C#
C++/Python

RGB/Depth/Segm. views

Yes

Yes

Yes

Multi-agent control

Yes

Yes

Yes

ROS interface

ROS2/Python

ROS/C++

ROS/C++

Hardware-In-The-Loop

No

Yes

No

Fully steppable physics

Yes

No

Yes

Aerodynamic effects

Drag, downwash, ground

Drag

Drag

OpenAI Gym (`https://github.com/openai/gym/blob/master/gym/core.py`) interface

Yes

Yes (https://github.com/microsoft/AirSim/pull/3215)

Yes

RLlib MultiAgentEnv (https://github.com/ray-project/ray/blob/master/rllib/env/multi_agent_env.py) interface

Yes

No

No

Performance

Simulation speed-up with respect to the wall clock when using

- *240Hz* (in simulation clock) PyBullet physics for **EACH** drone

- **AND** *48Hz* (in simulation clock) proportional–integral–derivative controller (PID) control of **EACH** drone

- **AND** nearby *obstacles* **AND** a mildly complex *background* (see GIFs)

- **AND** *24FPS* (in sim. clock), *64x48 pixel* capture of *6 channels* (RGBA, depth, segm.) on **EACH** drone

Rendering

OpenGL

CPU-based TinyRenderer

Single drone, **no** vision

15.5x

16.8x

Single drone **with** vision

10.8x

1.3x

Multi-drone (10), **no** vision

2.1x

2.3x

Multi-drone (5) **with** vision

2.5x

0.2x

80 drones in 4 env, **no** vision

0.8x

0.95x

Note   use gui=False and aggregate_phy_steps=int(SIM_
HZ/CTRL_HZ) for better performance.

While it is easy to—consciously or not—cherry pick (https://
en.wikipedia.org/wiki/Cherry_picking) statistics, ~5kHz PyBullet
physics (CPU-only) is faster than AirSim (1kHz) (https://arxiv.org/
pdf/1705.05065.pdf) and more accurate than Flightmare's 35kHz simple
single-quadcopter dynamics (https://arxiv.org/pdf/2009.00563.pdf).

Exploiting parallel computation—i.e., multiple (80) drones in
multiple (4) environments see script parallelism.sh (https://github.
com/utiasDSL/gym-pybullet-drones/blob/master/experiments/
performance/parallelism.sh)—achieves PyBullet physics updates at
~20kHz. Multi-agent six-channel video capture at ~750kB/s with CPU
rendering ((64*48)*(4+4+2)*24*5*0.2) is comparable to Flightmare's 240
RGB frames/s (https://arxiv.org/pdf/2009.00563.pdf) ((32*32)*3*240)
and is up to an order of magnitude faster on Ubuntu, with OpenGL
rendering.

Requirements and Installation

The repo was written using *Python 3.7* with conda (https://github.com/
JacopoPan/a-minimalist-guide) on *macOS 10.15* and tested on
macOS 11, Ubuntu 18.04.

On macOS and Ubuntu

Major dependencies are gym (`https://gym.openai.com/docs/`), pybullet (`https://docs.google.com/document/d/1OsXEhzFRSnvFcl3XxNGhn D4N2SedqwdAvK3dsihxVUA/edit`), stable-baselines3 (`https://stable-baselines3.readthedocs.io/en/master/guide/quickstart.html`), and rllib (`https://docs.ray.io/en/master/rllib.html`).

```
pip3 install --upgrade numpy Pillow matplotlib cycler
pip3 install --upgrade gym pybullet stable_baselines3 'ray[rllib]'
```

Video recording requires you to have ffmpeg (`https://ffmpeg.org/`) installed on *macOS*.

```
$ brew install ffmpeg
```

On *Ubuntu*:

```
$ sudo apt install ffmpeg
```

The repo is structured as a gym environment (`https://github.com/openai/gym/blob/master/docs/creating-environments.md`) and can be installed with `pip install –editable`:

```
$ git clone https://github.com/utiasDSL/gym-pybullet-drones.git
$ cd gym-pybullet-drones/
$ pip3 install -e .
```

On Ubuntu and with a GPU available, optionally uncomment line 203 (`https://github.com/utiasDSL/gym-pybullet-drones/blob/fab619b119e7deb6079a292a04be04d37249d08c/gym_pybullet_drones/envs/BaseAviary.py#L203`) of BaseAviary.py to use the eglPlugin (`https://support.google.com/drive/answer/6283888#heading=h.778da594xyte`).

On Windows

Check out these step-by-step <u>instructions</u> written by Dr. Karime Pereida for *Windows 10*.

Examples

There are two basic template scripts in examples/: fly.py and learn.py.

fly.py runs an independent flight using PID control implemented in class DSLPIDControl (https://github.com/utiasDSL/gym-pybullet-drones/tree/master/gym_pybullet_drones/control/DSLPIDControl.py):

```
$ cd gym-pybullet-drones/examples/
$ python3 fly.py # Try 'python3 fly.py -h' to show the script's
customizable parameters
```

Tip   Use the GUI's sliders and buttons. Use GUI RPM to override the control with interactive inputs.

---

learn.py is an RL example to learn take-off using stable-baselines3's A2C (https://stable-baselines3.readthedocs.io/en/master/modules/a2c.html) or rllib's <u>PPO</u>:

```
$ cd gym-pybullet-drones/examples/
$ python3 learn.py # Try 'python3 learn.py -h' to show the
script's customizable parameters
```

There are other scripts in the examples folder.

downwash.py is a flight script with only two drones, to test the downwash model:

```
$ cd gym-pybullet-drones/examples/
$ python3 downwash.py # Try 'python3 downwash.py -h' to show
the script's customizable parameters
```

compare.py replays and compares to a trace saved in example_trace.pkl (https://github.com/utiasDSL/gym-pybullet-drones/tree/master/ files/example_trace.pkl):

```
$ cd gym-pybullet-drones/examples/
$ python3 compare.py # Try 'python3 compare.py -h' to show the
script's customizable parameters
```

Experiments

Folder experiments/learning (https://github.com/utiasDSL/gym-pybullet-drones/tree/master/experiments/learning) contains scripts with template learning pipelines. For single-agent RL problems, using stable-baselines3 (https://stable-baselines3.readthedocs.io/en/master/guide/quickstart.html), run the training script (https://github.com/utiasDSL/gym-pybullet-drones/blob/master/experiments/learning/singleagent.py) as follows:

```
$ cd gym-pybullet-drones/experiments/learning/
$ python3 singleagent.py --env <env> --algo <alg> --obs
<ObservationType> --act <ActionType> --cpu <cpu_num>
```

Run the replay script (https://github.com/utiasDSL/gym-pybullet-drones/blob/master/experiments/learning/test_singleagent.py) to visualize the best trained agent(s):

```
$ python3 test_singleagent.py --exp ./results/save-<env>-
<algo>-<obs>-<act>-<time-date>
```

For multi-agent RL problems, using rllib (https://docs.ray.io/en/ master/rllib.html) run the train script (https://github.com/utiasDSL/ gym-pybullet-drones/blob/master/experiments/learning/ multiagent.py):

```
$ cd gym-pybullet-drones/experiments/learning/
$ python3 multiagent.py --num_drones <num_drones> --env <env>
--obs <ObservationType> --act <ActionType> --algo <alg> --num_
workers <num_workers>
```

Run the replay script (https://github.com/utiasDSL/gym-pybullet-drones/blob/master/experiments/learning/test_multiagent.py) to visualize the best trained agent(s):

```
$ python3 test_multiagent.py --exp ./results/save-<env>-<num_
drones>-<algo>-<obs>-<act>-<date>
```

Class BaseAviary

A flight arena for one (or more) quadrotor can be created as a subclass of BaseAviary():

```
>>> env = BaseAviary(
>>> drone_model=DroneModel.CF2X, # See DroneModel Enum class
for other quadcopter models
>>> num_drones=1, # Number of drones
>>> neighborhood_radius=np.inf, # Distance at which drones are
considered neighbors, only used for multiple drones
>>> initial_xyzs=None, # Initial XYZ positions of the drones
>>> initial_rpys=None, # Initial roll, pitch, and yaw of the
drones in radians
>>> physics: Physics=Physics.PYB, # Choice of (PyBullet)
physics implementation
```

```
>>> freq=240, # Stepping frequency of the simulation
>>> aggregate_phy_steps=1, # Number of physics updates within
each call to BaseAviary.step()
>>> gui=True, # Whether to display PyBullet's GUI, only use
this for debuging
>>> record=False, # Whether to save a .mp4 video (if gui=True)
or .png frames (if gui=False) in gym-pybullet-drones/files/,
see script /files/videos/ffmpeg_png2mp4.sh for encoding
>>> obstacles=False, # Whether to add obstacles to the
environment
>>> user_debug_gui=True) # Whether to use addUserDebugLine and
addUserDebugParameter calls (it can slow down the GUI)
```

And instantiated with gym.make()—see <u>learn.py</u> for an example

```
>>> env = gym.make('rl-takeoff-aviary-v0') # See learn.py
```

Then, the environment can be stepped with

```
>>> obs = env.reset()
>>> for _ in range(10*240):
>>> obs, reward, done, info = env.step(env.action_space.
sample())
>>> env.render()
>>> if done: obs = env.reset()
>>> env.close()
```

Creating New Aviaries

A new RL problem can be created as a subclass of BaseAviary
(https://github.com/utiasDSL/gym-pybullet-drones/blob/
master/gym_pybullet_drones/envs/BaseAviary.py) (i.e., class
NewAviary(BaseAviary): ...) and by implementing the following seven
abstract methods:

```
>>> #### 1
>>> def _actionSpace(self):
>>> # e.g., return spaces.Box(low=np.zeros(4), high=np.ones(4),
dtype=np.float32)
>>> #### 2
>>> def _observationSpace(self):
>>> # e.g., return spaces.Box(low=np.zeros(20), high=np.ones(20),
dtype=np.float32)
>>> #### 3
>>> def _computeObs(self):
>>> # e.g., return self._getDroneStateVector(0)
>>> #### 4
>>> def _preprocessAction(self, action):
>>> # e.g., return np.clip(action, 0, 1)
>>> #### 5
>>> def _computeReward(self):
>>> # e.g., return -1
>>> #### 6
>>> def _computeDone(self):
>>> # e.g., return False
>>> #### 7
>>> def _computeInfo(self):
>>> # e.g., return {"answer": 42} # Calculated by the Deep Thought
supercomputer in 7.5M years
```

See CtrlAviary (https://github.com/utiasDSL/gym-pybullet-drones/blob/master/gym_pybullet_drones/envs/CtrlAviary.py), VisionAviary (https://github.com/utiasDSL/gym-pybullet-drones/blob/master/gym_pybullet_drones/envs/VisionAviary.py), HoverAviary (https://github.com/utiasDSL/gym-pybullet-drones/blob/master/gym_pybullet_drones/envs/single_agent_rl/HoverAviary.py), and FlockAviary (https://github.com/utiasDSL/gym-pybullet-drones/blob/master/gym_pybullet_drones/envs/multi_agent_rl/FlockAviary.py) for examples.

Action Spaces Examples

The action space's definition of an environment must be implemented in each subclass of BaseAviary (https://github.com/utiasDSL/gym-pybullet-drones/blob/master/gym_pybullet_drones/envs/BaseAviary.py) by function.

```
>>> def _actionSpace(self):
>>> ...
```

In CtrlAviary (https://github.com/utiasDSL/gym-pybullet-drones/blob/master/gym_pybullet_drones/envs/CtrlAviary.py) and VisionAviary (https://github.com/utiasDSL/gym-pybullet-drones/blob/master/gym_pybullet_drones/envs/VisionAviary.py), it is a Dict() (https://github.com/openai/gym/blob/master/gym/spaces/dict.py) of Box(4,) (https://github.com/openai/gym/blob/master/gym/spaces/box.py) containing the drones' commanded RPMs.

The dictionary's keys are "0", "1", .., "n"—where n is the number of drones.

Each subclass of BaseAviary (https://github.com/utiasDSL/gym-pybullet-drones/blob/master/gym_pybullet_drones/envs/BaseAviary.py) also needs to implement a preprocessing step translating actions into RPMs:

```
>>> def _preprocessAction(self, action):
>>> ...
```

CtrlAviary (https://github.com/utiasDSL/gym-pybullet-drones/blob/master/gym_pybullet_drones/envs/CtrlAviary.py), VisionAviary (https://github.com/utiasDSL/gym-pybullet-drones/blob/master/gym_pybullet_drones/envs/VisionAviary.py), HoverAviary (https://github.com/utiasDSL/gym-pybullet-drones/blob/master/gym_pybullet_drones/envs/single_agent_rl/HoverAviary.py), and FlockAviary (`https://github.com/utiasDSL/gym-pybullet-drones/blob/master/gym_pybullet_drones/envs/multi_agent_rl/FlockAviary.py`) all simply clip the inputs to `MAX_RPM`.

DynAviary's action input to DynAviary.step() is a Dict() of Box(4,) containing (`https://github.com/utiasDSL/gym-pybullet-drones/blob/master/gym_pybullet_drones/envs/DynAviary.py`) the following:

- The desired thrust along the drone's z-axis

- The desired torque around the drone's x-axis

- The desired torque around the drone's y-axis

- The desired torque around the drone's z-axis

From these, desired RPMs are computed by DynAviary._preprocessAction() (`https://github.com/utiasDSL/gym-pybullet-drones/blob/master/gym_pybullet_drones/envs/DynAviary.py`).

Observation Spaces Examples

The observation space's definition of an environment must be implemented by every subclass of BaseAviary (`https://github.com/utiasDSL/gym-pybullet-drones/blob/master/gym_pybullet_drones/envs/BaseAviary.py`):

```
>>> def _observationSpace(self):
>>> ...
```

CtrlAviary (`https://github.com/utiasDSL/gym-pybullet-drones/blob/master/gym_pybullet_drones/envs/CtrlAviary.py`) is a Dict() (`https://github.com/openai/gym/blob/master/gym/spaces/dict.py`) of pairs {"state": Box(20,), "neighbors": MultiBinary(num_drones)}.

The dictionary's keys are "0", "1", .., "n"—where *n* is the number of drones.

Each Box(20,) (`https://github.com/openai/gym/blob/master/gym/spaces/box.py`) contains the drone's

- X, Y, Z position in `WORLD_FRAME` (in meters, 3 values)

- Quaternion orientation in `WORLD_FRAME` (4 values)

- Roll, pitch, and yaw angles in `WORLD_FRAME` (in radians, 3 values)

- The velocity vector in `WORLD_FRAME` (in m/s, 3 values)

- Angular velocity in `WORLD_FRAME` (3 values)

- Motors' speeds (in RPMs, 4 values)

Each MultiBinary(num_drones) (`https://github.com/openai/gym/blob/master/gym/spaces/multi_binary.py`) contains the drone's own row of the multi-robot system adjacency matrix (`https://en.wikipedia.org/wiki/Adjacency_matrix`).

The observation space of VisionAviary (`https://github.com/utiasDSL/gym-pybullet-drones/blob/master/gym_pybullet_drones/envs/VisionAviary.py`) is the same as that for CtrlAviary (`https://github.com/utiasDSL/gym-pybullet-drones/blob/master/gym_pybullet_drones/envs/CtrlAviary.py`) but also includes keys rgb, dep, and seg (in each drone's dictionary) for the matrices containing the drone's RGB, depth, and segmentation views.

To fill/customize the content of obs, every subclass of BaseAviary (`https://github.com/utiasDSL/gym-pybullet-drones/blob/master/gym_pybullet_drones/envs/BaseAviary.py`) needs to implement the following:

```
>>> def _computeObs(self, action):
>>> ...
```

See BaseAviary._exportImage() (`https://github.com/utiasDSL/`
`gym-pybullet-drones/blob/master/gym_pybullet_drones/envs/`
`BaseAviary.py`) and its use in VisionAviary._computeObs() (`https://`
`github.com/utiasDSL/gym-pybullet-drones/blob/master/gym_`
`pybullet_drones/envs/VisionAviary.py`) to save frames as PNGs.

Obstacles

Objects can be added to an environment using loadURDF (or loadSDF,
loadMJCF) in method _addObstacles()

```
>>> def _addObstacles(self):
>>> ...
>>> p.loadURDF("sphere2.urdf", [0,0,0], p.getQuaternionFromEul
er([0,0,0]), physicsClientId=self.CLIENT)
```

Drag, Ground Effect, and Downwash Models

Simple drag, ground effect, and downwash models can be included in
the simulation by initializing BaseAviary() with physics=Physics.
PYB_GND_DRAG_DW; these are based on the system identification of Forster
(2015) (`http://mikehamer.info/assets/papers/Crazyflie Modelling.`
`pdf`) (Eq. 4.2), the analytical model used as a baseline for comparison by
Shi et al. (2019) (`https://arxiv.org/pdf/1811.08027.pdf`) (Eq. 15), and
DSL's (`https://www.dynsyslab.org/vision-news/`) experimental work.

Check the implementations of _drag(), _groundEffect(), and
_downwash() in BaseAviary (`https://github.com/utiasDSL/gym-`
`pybullet-drones/blob/master/gym_pybullet_drones/envs/`
`BaseAviary.py`) for more detail.

PID Control

Folder control (https://github.com/utiasDSL/gym-pybullet-drones/blob/master/gym_pybullet_drones/control/) contains the implementations of two PID controllers: DSLPIDControl (https://github.com/utiasDSL/gym-pybullet-drones/blob/master/gym_pybullet_drones/control/DSLPIDControl.py) (for DroneModel.CF2X/P) and SimplePIDControl (https://github.com/utiasDSL/gym-pybullet-drones/blob/master/gym_pybullet_drones/control/SimplePIDControl.py) (for DroneModel.HB). They can be used as follows:

```
>>> ctrl = [DSLPIDControl(drone_model=DroneModel.CF2X) for i in
range(num_drones)] # Initialize "num_drones" controllers
>>> ...
>>> for i in range(num_drones): # Compute control for
each drone
>>> action[str(i)], _, _ = ctrl[i].computeControlFromState(. #
Write the action in a dictionary
>>> control_timestep=env.TIMESTEP,
>>> state=obs[str(i)]["state"],
>>> target_pos=TARGET_POS)
```

For high-level coordination—using a *velocity input*—VelocityAviary (https://github.com/utiasDSL/gym-pybullet-drones/blob/master/gym_pybullet_drones/envs/VelocityAviary.py) integrates PID control within a gym.Env.

The setPIDCoefficients (https://github.com/utiasDSL/gym-pybullet-drones/blob/master/gym_pybullet_drones/control/BaseControl.py) method can be used to change the coefficients of one of the given PID controllers—and, for example, to implement learning problems whose goal is parameter tuning (see TuneAviary (https://github.com/utiasDSL/gym-pybullet-drones/blob/master/gym_pybullet_drones/envs/single_agent_rl/TuneAviary.py)).

Logger

Class Logger (https://github.com/utiasDSL/gym-pybullet-drones/blob/master/gym_pybullet_drones/utils/Logger.py) contains helper functions to save and plot simulation data, as in this example:

```
>>> logger = Logger(logging_freq_hz=freq, num_drones=num_
drones) # Initialize the logger
>>> ...
>>> for i in range(NUM_DRONES): # Log information for
each drone
>>> logger.log(drone=i,
>>> timestamp=K/env.SIM_FREQ,
>>> state= obs[str(i)]["state"],
>>> control=np.hstack([ TARGET_POS, np.zeros(9) ]))
>>> ...
>>> logger.save() # Save data to file
>>> logger.plot() # Plot data
```

ROS2 Python Wrapper

Workspace ros2 (https://github.com/utiasDSL/gym-pybullet-drones/tree/master/ros2) contains two ROS2 Foxy Fitzroy (https://index.ros.org/doc/ros2/Installation/Foxy/) Python nodes:

- AviaryWrapper (https://github.com/utiasDSL/gym-pybullet-drones/blob/master/ros2/src/ros2_gym_pybullet_drones/ros2_gym_pybullet_drones/aviary_wrapper.py) is a wrapper node for a single-drone CtrlAviary (https://github.com/utiasDSL/gym-pybullet-drones/blob/master/gym_pybullet_drones/envs/CtrlAviary.py) environment.

- RandomControl (https://github.com/utiasDSL/
  gym-pybullet-drones/blob/master/ros2/src/ros2_
  gym_pybullet_drones/ros2_gym_pybullet_drones/
  random_control.py) reads AviaryWrapper's obs topic
  and publishes random RPMs on topic.

With ROS2 installed (on either macOS or Ubuntu, edit ros2_and_pkg_
setups.(zsh/bash) accordingly), run the following:

```
$ cd gym-pybullet-drones/ros2/
$ source ros2_and_pkg_setups.zsh # On macOS, on Ubuntu use
$ source ros2_and_pkg_setups.bash
$ colcon build --packages-select ros2_gym_pybullet_drones
$ source ros2_and_pkg_setups.zsh # On macOS, on Ubuntu use
$ source ros2_and_pkg_setups.bash
$ ros2 run ros2_gym_pybullet_drones aviary_wrapper
In a new terminal terminal, run
$ cd gym-pybullet-drones/ros2/
$ source ros2_and_pkg_setups.zsh # On macOS, on Ubuntu use
$ source ros2_and_pkg_setups.bash
$ ros2 run ros2_gym_pybullet_drones random_control
```

Desiderata/WIP

- Template scripts using PyMARL
  (https://github.com/oxwhirl/pymarl)

- Google Collaboratory (https://colab.research.
  google.com/notebooks/intro.ipynb) example

- Alternative multi-contribution downwash effect

Citation

If you wish, please cite our work link (`https://arxiv.org/abs/2103.02142`) as

```
@INPROCEEDINGS{panerati2021learning,
    title={Learning to Fly---a Gym Environment with PyBullet
    Physics for Reinforcement Learning of Multi-agent
    Quadcopter Control},
    author={Jacopo Panerati and Hehui Zheng and SiQi Zhou and
    James Xu and Amanda Prorok and Angela P. Schoellig},
    booktitle={2021 IEEE/RSJ International Conference on
    Intelligent Robots and Systems (IROS)},
    year={2021},
    volume={},
    number={},
    pages={},
    doi={}
}
```

References

Michael, Nathan, Daniel Mellinger, Quentin Lindsey, and Vijay Kumar. (2010). *The GRASP Multiple Micro UAV Testbed* (`http://citeseerx.ist.psu.edu/viewdoc/download?doi=10.1.1.169.1687&rep=rep1&type=pdf`).

Landry, Benoit. (2014). *Planning and Control for Quadrotor Flight through Cluttered Environments* (`http://groups.csail.mit.edu/robotics-center/public_papers/Landry15`)

Forster, Julian. (2015). *System Identification of the Crazyflie 2.0 Nano Quadrocopter* (`http://mikehamer.info/assets/papers/Crazyflie Modelling.pdf`)

Luis, Carlos, and Jeroome Le Ny. (2016). *Design of a Trajectory Tracking Controller for a Nanoquadcopter* (https://arxiv.org/pdf/1608.05786.pdf)

Shah, Shital, Debadeepta Dey, Chris Lovett, and Ashish Kapoor. (2017). *AirSim: High-Fidelity Visual and Physical Simulation for Autonomous Vehicles* (https://arxiv.org/pdf/1705.05065.pdf)

Liang, Eric, Richard Liaw, Philipp Moritz, Robert Nishihara, Roy Fox, Ken Goldberg, Joseph E. Gonzalez, Michael I. Jordan, and Ion Stoica. (2018). *RLlib: Abstractions for Distributed Reinforcement Learning* (https://arxiv.org/pdf/1712.09381.pdf)

Raffin, Antonin, Ashley Hill, Maximilian Ernestus, Adam Gleave, Anssi Kanervisto, and Noah Dormann. (2019) *Stable Baselines 3* (https://github.com/DLR-RM/stable-baselines3)

Shi, Guanya, Xichen Shi, Michael O'Connell, Rose Yu, Kamyar Azizzadenesheli, Animashree Anandkumar, Yisong Yue, and Soon-Jo Chung. (2019). *Neural Lander: Stable Drone Landing Control Using Learned Dynamics* (https://arxiv.org/pdf/1811.08027.pdf)

Samvelyan, Mikayel, Tabish Rashid, Christian Schroeder de Witt, Gregory Farquhar, Nantas Nardelli, Tim G. J. Rudner, Chia-Man Hung, Philip H. S. Torr, Jakob Foerster, and Shimon Whiteson. (2019). *The StarCraft Multi-Agent Challenge* (https://arxiv.org/pdf/1902.04043.pdf)

Liu, C. Karen, and Dan Negrut. (2020). *The Role of Physics-Based Simulators in Robotics* (https://www.annualreviews.org/doi/pdf/10.1146/annurev-control-072220-093055)

Song, Yunlong, Selim Naji, Elia Kaufmann, Antonio Loquercio, and Davide Scaramuzza. (2020). *Flightmare: A Flexible Quadrotor Simulator* (https://arxiv.org/pdf/2009.00563.pdf)

Bonus GIF for scrolling this far

University of Toronto's Dynamic Systems Lab (`https://github.com/utiasDSL`) / Vector Institute(`https://github.com/VectorInstitute`) / University of Cambridge's Prorok Lab (`https://github.com/proroklab`) / Mitacs (`https://www.mitacs.ca/en/projects/multi-agent-reinforcement-learning-decentralized-uavugv-cooperative-exploration`)

APPENDIX C

Introduction to the Future of AI & ML Research

Research that seeks to revolutionize artificial intelligence (AI) and machine learning (ML) needs to occur. AI and ML have been essentially stagnant since the inception of the convolutional neural network and deep learning in 1994 by Yann LeCun, even though they were improved upon in 1998 with the advent of the LeNet 5 (LeCun, Bottou, Bengio, & Haffner, 1998; LeCun, 2015). While small advances have occurred over the past 11 years in both hardware and software, the fact remains that for the most part artificial intelligence has been trapped using "traditional" statistical learning and pattern recognition to draw inferences and predict for the past 15 years. To advance AI and ML, the field has to break through this barrier and enter the "Third Wave" of AI (DARPA, 2018), which will require innovative upheaval. The Third Wave consists of systems that construct contextual explanatory models for classes of real-world phenomena (DARPA/I20, 2018). This will open the door to solving the unexpected query (UQ) problem.

© David Allen Blubaugh, Steven D. Harbour, Benjamin Sears, Michael J. Findler 2022
D. A. Blubaugh et al., *Intelligent Autonomous Drones with Cognitive Deep Learning*,
https://doi.org/10.1007/978-1-4842-6803-2

Third Wave of AI

"Facing data bandwidth constraints and ever-rising computational requirements, sensing and computing must reinvent themselves by mimicking neurobiological architectures (and processes), claimed a recently published report by Yole Développement (Lyon, France)" (2019, `https://www.embedded.com/making-the-case-for-neuromorphic-chips-for-ai-computing/`). Hence, the rise of neuromorphic computing, spiking neural networks, and, most recently, advances in reinforcement learning research (Patel, Hazan, Saunders, Siegelmann, & Kozma, 2019). The robustness of reinforcement learning policies improves upon conversion to spiking neuronal network platforms. However, it is quite clear that this research must include exploration of the human brain and mind, as well (Stöckel & Eliasmith, 2019; Applied Brain Research, 2019). Consequently, emphasis has been placed on "mimicking neurobiological architectures (and processes)", and it should be noted that this also implies the inclusion of neurocognitive processes because, without the inclusion of neuroscience research, this revolution into Third Wave AI is not fully attainable. It is vital to research in Spiking Neural Networks (SNNs), Reinforcement Learning (REL), and neuromorphic computing that rigorous research into neuroscience must also occur, according to Demis Hassabis, neuroscientist and co-founder of Deep Mind (2019). Learning how our brain works—from the neuron cell level up to and including the macro cognitive thought processes level—must also occur, which can be done by utilizing neuroergonomics. This is analogous to type I and type II processes in current ToM (Stanovich, 2015).

Contextual adaptation engineers create systems that construct explanatory models for classes of real-world phenomena. AI systems learn and reason as they encounter new tasks and situations and as natural communication among machines and people happens.

To do this, experiments must be set up that utilize electroencephalograms, magnetic resonance imaging, and functional infrared imaging of the brain while integrated with AI and ML, all while a person is performing tasks (consciously and/or subconsciously). Consequently, this type of equipment would have to be purchased.

Future Third Wave AI technologies include human–machine symbiosis, an automatic whole-system causal model, continuous learning, embedded machine learning, and explainable AI.

Index

A

© David Allen Blubaugh, Steven D. Harbour, Benjamin Sears, Michael J. Findler 2022   495
D. A. Blubaugh et al., *Intelligent Autonomous Drones with Cognitive Deep Learning*,
https://doi.org/10.1007/978-1-4842-6803-2

T

Printed in the United States
by Baker & Taylor Publisher Services